Prospects for Tomorrow

The Future of Food

Brian J. Ford

Thames & Hudson

First published in the United Kingdom in 2000 by Thames & Hudson Ltd,
181A High Holborn, London WC1V 7QX

British Library Cataloguing-in-Publication Data
A catalogue record for this book is available from the British Library

ISBN 0-500-28075-4

Printed and bound in Slovenia by Mladinska Knjiga

CONTENTS

INTRODUCTION

When we plan for the future, food will pose some of the greatest problems. Our daily diet is rapidly changing, new food-borne diseases are appearing and ideas on safety are in a state of flux. In Britain 55 per cent of our meals are prepared outside the home, in the USA 75 per cent, and yet the amount we spend on food is falling. In the 1950s families spent a quarter of their income on food; by 2000 it was an eighth.

We now produce enough food for everyone, yet a third of the world is still hungry. We are faced with global climate change, mostly caused by the industrialized world, yet threatening some of the most populous nations. Organic farming is becoming more popular among the wealthier nations, while genetically modified crops – heralded as a great leap forward – have been widely rejected. Monsanto, whose research attracted so much opprobrium in 1999, found by 2000 that GMOs had been banished from their own kitchens by their caterers. The future of food is a matter of dispute that will affect us all.

Many friends in Europe, the Middle and Far East and the Americas have helped me over the years. My lectures as Vice-Chairman of Nutrition at the Royal Society of Health and at the Mosimann Academy in London have brought me into contact with many learned minds, and I am grateful to Doug Kentish, executive producer of our TV series *Food for Thought*, for complete immersion in the food industry.

Drafts of this book were reviewed by Professor Alan Malcolm, chief executive of the Institute of Biology, where it has been my privilege to serve as a Member of Council, by Dr John S. Slade, former chief virologist at the Thames Water Authority, and Professor David Conning, former director general of the British Nutrition Foundation. To them, to my family for all their help and to the many friends around the world who have taught me about food and culture, I am very grateful.

Chapter 1

What Do We Eat?

This book is for everyone who eats food. Of all the people I have met on my travels only one had never bothered with it.

Food strengthens the body, energizes our actions, employs farmers, boosts the bureaucracy of governments and makes mighty giants of supermarkets who overcharge the public. Food kills incalculable numbers of people every year, spreads epidemics, destroys landscapes and cultures, and clogs our arteries. It perpetrates cruelty and creates contentment, pacifies or enrages children and eases social interaction.

For centuries people were content to exist on a diet that was slow to change. In recent decades that has altered: new products (like hamburgers and hot dogs) have spread across the world. People whose parents regarded curry as a craze regularly patronize Indian restaurants. Oriental cultures have spread eateries from Korea, Japan and Vietnam across the nations who fought deadly wars against them. New international dishes like chop suey, tikka marsala and chicken nuggets have been created, and the creations of working people, from Cornish pasties to baked beans, have gained a broad popularity.

School meals have changed out of recognition within a generation. Pizza and lasagne, once confined to Italy, are now major commodities across the developed world. This book steps warily into the future, looking at the extraordinary changes we can expect to see and presenting some of the novel foods on which our descendants will depend.

The Basic Diet

Our diet will change dramatically in the future, although the essential components that we need to eat in order to stay healthy remain the same. We evolved to subsist on a mixed intake of foodstuffs, and a traditional varied diet provides everything we need. Problems arise because we now eat a diet where the components are chosen by producers, so we can now be attracted to foods that are rich in sugar or salt, or contain far more saturated fats than anything for which we evolved.

To stay alive and grow, the main groups of components needed are:

- Fats – our cells depend upon oily and fatty compounds to function properly. Cholesterol, an excess of which can block blood vessels, is actually a vital component of every living cell. Fats are not inherently 'bad'. Some recent research has given us synthetic fats which are not utilized by the body as an energy store. How they could ever be removed from the tissues (once deposits had been laid down) remains a mystery.
- Carbohydrates – the starches and sugars in our diets which provide the energy for life. These compounds can burn in air with a flame. Within our bodies, they combine with oxygen just as they do when they burn, but as part of the cells' chemistry. They release heat energy that helps regulate our inner temperature.
- Proteins – made from amino acids. Proteins are the substances that make up the living matter of all cells. The diet must contain a suitable range of proteins and amino acids. Human beings developed for an omnivorous diet, eating both animal and vegetable foods, but a purely vegetarian diet is possible if we take care that all the essential amino acids are present in our food.
- Minerals – like cobalt and chromium, iron and magnesium, are involved in the complex chemistry that keeps us alive. Cows eating spring grass may consume too little magnesium, for instance. They collapse and can die within hours – but a single injection of magnesium salts provides an immediate return to full health. Calcium is important for the making of teeth and bones. They probably evolved as a reservoir for extra calcium, and can become weak and spongy if too much calcium returns from the bones to the bloodstream.
- Vitamins – first recognized by Sir Frederick Hopkins in England in 1906, when he discovered that there were some crucial accessory factors that food must contain for a healthy diet. Casimir Funk, a Polish biochemist, coined the term vitamines in 1912. The existence of vitamin deficiency had been recorded long before vitamins were recognized, and successful treatment had already been introduced.

TABLE OF VITAMINS

	Name	Identity	Soluble in	Source	Effects
A	Retinol	alpha-and beta-carotene	fat	butter, fish oil	vital for sight; excess causes hair and skin loss, nausea
B_1	Thiamine	thiamine pryophosphate and carcoboxylase	water	cereals, pork	aids nerve function; lack causes beriberi
B_2	Riboflavin	flavin mononucleotide and flavin adenine dinucleotide	water	liver, dairy produce	vital for cell membranes, tongue, skin
B_3	Niacin	nicotinamide (nicotinic acid)	water	meat	prevents pellagra
B_6	Pyridoxine	pyridoxal phosphate or codecarboxylase	water	meat, nuts	important for blood and nerve transmitters
B_{12}	Folic acid	pterolic acids and their esters and cobamide coenzymes	water	egg, meat, milk	lack causes pernicious anaemia
	Coenzyme A	pantothenic acid	water	offal, yeast	none known
B_{12a}	Aquocobalamin				
B_{12b}	Hydroxycobalamin		water	offal, egg, fish	vital for DNA synthesis
B_{12c}	Nitrocobalamin				
B_{13}	Not yet confirmed				
B_{15}	Not yet confirmed				
C	Ascorbic acid		water	fruit, parsley, peppers	prevents scurvy
D	Calciferol	ergosterol, 7-dehydrocholesterol	fat	action of sunlight in skin, milk	lack causes rickets
E	Tocopherol	alpha-, beta-, gamma-, delta-, epsilon-, and eta-tocopherol	fat	plant oils, leaves	antioxidants promote health
H	Biotin	1-N carboxybiotin-	water	egg, liver, yeast	essential for metabolism
K_1		phylloquinone	fat	green vegetables	vital for blood coagulation
$K_{2\,(20)}$		menaquinone-4	fat	—	—
$K_{2\,(30)}$		menaquinone-6	fat	—	—
K_3	Menadione	menaquinone	fat	—	—

Beriberi was the first vitamin-deficiency disease to be recorded. The term means extreme weakness in the Sinhalese language of Sri Lanka, and it is due to a lack of vitamin B. It was first documented by Chinese physicians in 2600 BC. Scurvy (due to a lack of vitamin C) was described in Ancient Egypt around 1500 BC. The first treatment of a vitamin deficiency was recorded by the French explorer Jacques Cartier, who sailed for North America in 1534. His crews were ravaged by scurvy, and Indian tribes showed him how to use an infusion of pine needles to help control the disease. The English physician James Lind finally showed that citrus fruits were rich in the vital ingredient, and during the voyages of Captain James Cook between 1768 and 1771 supplies of limes were issued to the men. These were the first long voyages that were not marked by the scourge of scurvy. The use of these fruit gave rise to the term 'limey', the nickname by which the English have long been known. Beriberi was first successfully treated in 1885, when a doctor in Japan supplemented a traditional diet of polished rice with meat and vegetables. Sufferers from beriberi dramatically improved, once their diet contained these extra sources of vitamin B.

Another vitamin in the B group, folic acid, has recently been associated with the campaign against neural tube defects, such as spina bifida. The claims that folic acid can end this scourge may be premature. Publicity encouraging people to take the vitamin has increased substantially. The Medical Expert Advisory Group in the United Kingdom recommended in 1992 that women who were trying to conceive should take 0.4 mg folic acid each day, and these recommendations were soon headline news in all the magazines. Folic acid featured on TV, and was splashed across the newspapers. All the chemists' shops stocked up, ready for the stampede.

The statistics were initially very impressive. In the early 1970s there were more than 200 cases of spina bifida out of every 100,000; it fell to less than 40, a drop of 80 per cent. What's interesting is that it fell by 1990, before folic acid was popular. A team in London, headed by Dr Rezan A. Kadir at the Royal Free and University College Medical School in London, inspected the figures and found – since folic acid has been widely available – the rate of decrease has actually slowed down, and not (as one might expect) increased. Over-the-counter sales of folic acid in chemists' shops went up dramatically between 1990 and 1994, but they actually declined in from 1995 to 1996. Thus, the incidence of spina bifida decreased most dramatically when folic acid was not widely purchased. When it was being bought in large amounts, the rate of improvement decreased. Whatever conclusions may be drawn, it is clear that simple notions of cause- and-effect do not always work where these vitamin supplements are concerned.

Similar controversies surround vitamin C, ascorbic acid. As we have seen, this compound is found in citrus fruits like lemons, limes and oranges, and also in vegetables ranging from sweet peppers to parsley. Ascorbic acid is important for the formation of collagen in the body, the fibrous material that helps hold cells together. A lack of ascorbic acid leads to a failure to maintain collagen, and the body can actually come apart. Old healed scars reopen, and the muscles of the heart lose their strength. Although we think of scurvy as a disease of former centuries, it is sometimes found in the Western world among old people who do not eat enough fruit.

More recently it has been said that ascorbic acid helps protect the body against colds and 'flu. Much publicity has been given to the idea that a gram of vitamin C every day helps to ward off such virus infections. It has to be said that repeated research trials have failed to prove the point. However, like a great many people, I always take some supplemental vitamin C tablets when the first symptoms of a cold appear, and have not suffered seriously from those infections for over a decade. It may be that some subtle relationship does, after all, exist.

Vitamin D, calciferol, is centrally involved in calcium metabolism and is necessary for the proper formation of teeth and bones. A deficiency of vitamin D results in rickets. Fatty fish, margarine and eggs provide a plentiful supply of calciferol, and it is also synthesized in our skin by the action of sunlight. In the UK signs of rickets can still be found among some immigrant ethnic groups who keep themselves swathed in robes.

The tocopherols (vitamin E) are found in plant material, and have long been associated with sexual prowess. They act as antioxidants, helping to mop up potentially dangerous free radicals in the body.

'Vitamin F' was the name given for a time to the prostaglandin precursors like gamma-linoleic acid, found in evening primrose oil. It is now downgraded to the vitamin-like compounds.

Vitamin K is the group of quinones that are abundant in leafy vegetables and beef liver, and, like the bioflavinoids once known as 'vitamin P_c,' are crucially important for successful blood coagulation.

The last true vitamin to be discovered was vitamin B_{12} in 1948. Its introduction led to the conquering of pernicious anaemia. Between 1929 and 1975 thirteen newly discovered vitamins were synthesized. Since then, research has continued into the benefits of vitamins in the diet. Attention was initially focused on their role in conquering deficiency diseases, but in recent years we have begun to investigate how they might delay the onset of degenerative diseases and cancer.

We now also recognize compounds which are similar to vitamins, and they are listed in the table below:

TABLE OF VITAMIN-LIKE COMPOUNDS

Name	Occurrence	Function
Choline	egg, liver, wheat germ, spinach	part of acetylcholine (neurotransmitter) and crucial for fat transport
Myoinositol	phospholipids	growth factor vital for yeasts and fungi
Para-aminobenzoic acid	part of folic acid molecule	vital for some microorganisms
Carnitine ('vitamin B_T')	synthesized in animals	essential for mealworms; may assist cardiac patients
Gamma-linoleic acid ('vitamin F')	evening primrose oil	prostaglandin precursor, may mollify ezcema
Lipoic acid	potatoes	antioxidant, essential for some microorganisms
Bioflavinoids ('vitamin P_c')	coloured fruit	may help to regulate control of bleeding

Now that we can produce vitamins industrially, they are readily available and cheap. They will become an important ingredient of tomorrow's food. Some foods (like breakfast cereals) have been fortified with vitamins for many years, and for the foods of the future the need will be to ensure a balanced intake of vitamins and other vital constituents for people with varying lifestyles and at every stage of life.

Rectifying vitamin deficiencies has been the main priority for the past half-century, but we can now look to the possible value of vitamins in larger than normal amounts. Some research has suggested that extra vitamins can help to maintain good health in the elderly, and there are many people who pop vitamins every day and swear by the benefits they confer. However, we have to take care to avoid overdosage. An excess of vitamin D can cause tiredness and weakness, loss of appetite and vomiting. Yellowish deposits appear in the skin and fingernails, and calcium deposits appear in organs like the kidneys. Young children consuming too much vitamin D fail to grow normally. Children face the greatest risk. A daily intake of 400 units of vitamin D is reckoned to be adequate for a child and children eating five times this dose can be damaged by the excess. In adults, effects appear when the dose is 50,000 units – so we need to bear in mind that children are much more sensitive to excesses of the vitamin than adults.

A raised intake of vitamins and other supplements may prove to help maintain health in the world of the future, but diets will have to be carefully planned if harmful excesses are to be avoided.

Deficient Diets

Everyone knows that deficient diets are found in developing countries, but it is surprising to realize that we can also find widespread evidence of a poor diet in the developed world. Many adults are fat because of the amount of saturated fat in the diet combined with a lack of exercise. We exercise far too little, and fast food can lead us into habits that leave important items, such as fresh fruit and vegetables and soluble fibre, out of our daily food intake. A lack of energy is not the problem: a lack of food quality, however, is.

During the last decade of the twentieth century, almost one person in ten of the global population had a diet with fewer calories than is needed for normal growth and activity. Sixteen per cent of the world's infants were born underweight and 38 per cent of children under five in the developing countries were underweight for their age. Twelve per cent of the people in the world suffered serious iron deficiency, and a quarter of the world's population was actually anaemic. Four per cent suffered mental deficiency due to a lack of iodine in the diet, many of them officially classified as cretins. Every year, as the millennium drew to a close, half a million children became permanently blind through a lack of vitamin A in their diet.

What will the future say of these scandalous figures? Essential dietary constituents like iron, iodine and vitamin A are cheap and readily available. A source of food that provides such essential components can easily be developed. One goal for the future must be to ensure that, whatever is said about the lack of food (see Chapter 6), the supply is of sufficient quality to help fight against the cruelty of deficiency diseases.

Throughout South America and in parts of Africa, malnutrition has not increased in recent decades. Indeed, a great swathe across the East, from Arabia through India and Pakistan to Indonesia and the Philippines, has shown steadily decreasing levels of malnutrition. Most of Africa, however, is witnessing a steady increase in dietary deficiencies, and this problem will become more severe as population levels continue to rise. We will examine the future of food supplies in Chapter 6, but we must understand that food quality has to come to the forefront of debate. Our current policies are creating a food supply that, even when it is abundant, is often far from healthy.

The Prophet Motive

Looking into the future is a dangerous practice. Futurologists imagine that they are being innovative, but most futuristic scenarios are either extrapolations of the present or objections to fashionable views. Most attempts are embarrassingly wide of the mark. Thomas Malthus, (1766–1834) was the English philosopher who in 1798 anonymously published *An Essay on the Principle of Population as it affects the Future Improvement of Society, with Remarks on the Speculations of Mr Godwin, Condorcet, and other Writers.* It began as a pamphlet, but by the time of the sixth edition in 1826 it had grown into a massive volume.

Malthus claimed that the growth in population would always outstrip food supply, until starvation and essential misery were the lot of humankind. He has become famous for his vision of global starvation, but in fact his book began as a young man's challenge to the views of his father, who, following Godwin and Condorcet, foresaw a glowing and optimistic future round the corner. Malthus's analysis was superficial, and it was also wrong. Population has increased much faster than he could have anticipated, yet the 1990s have been distinguished by our producing more than enough food to feed everyone, if only we could distribute it. The decades since Malthus's time have seen progress of a kind that flies in the face of everything he foresaw.

More recent, but equally influential was the report *Limits to Growth* from the Club of Rome, published in 1972. It, too, was essentially pessimistic. The events of the three decades since then have been far more complex, more responsive, much more unexpected and generally more positive than that report predicted. The control of population and the astonishing increase in agricultural productivity (notably in China) have shown how human technology can adapt. We have never been good at looking ahead: in the same year, 1972, the BBC produced a two-hour television documentary warning of the effects of emissions into the atmosphere. Accumulating smoke was threatening our planet with cooling, and the near future would condemn much of the industrialized world to a subterranean existence in malls that could protect us from the cold. Much of the discussion in the 1970s was preoccupied with global cooling, not warming.

Today's concerns about the greenhouse effect show how quickly our views can change. Throughout the 1970s there was a widespread belief that we were on the point of running out of food, energy and resources. Jet travel seemed doomed to become unendurably expensive; paper would run out; oil was in short supply; pollution would soon swamp us all. Large organizations suddenly began to circulate documents that were single-spaced, rather than double, and used both sides of the paper for the first time. They have reverted to their old habits since then.

Large cars soon fell dramatically in value. The public felt that the world had not long to survive. It was true that there was an oil crisis, though that was due to a lack of Middle Eastern goodwill rather than to a lack of oil in the ground. Within a decade we were back to our old ways of profligacy and profit, confident that the expansion would continue unabated. We produced more food, decade after decade, and saw the civilized nations becoming increasingly materialistic. Waste became a hallmark of consumerism. The small garbage bin of the 1960s was replaced with larger containers by the 1990s, and even they began to seem too small for a family's waste. Roads were clogged with juggernauts ferrying around vast quantities of consumer goods. It has been bonanza time.

In tomorrow's world, we will face serious new problems that we must address now. New oil reserves are steadily being discovered, but they are a finite resource, and their combustion is clearly altering the atmosphere. Currently we are consuming petrochemicals three or four times faster than we are discovering new reserves of crude oil, and the best estimates suggest that the level of oil production will peak around 2002. From then it likely to fall, and prices will rise. We have greatly increased food production, but still we fail to deliver it to those who need it most. Sound and solid holistic analysis of our problems, and international attempts to solve them, must be the hallmark of the new era.

The Death of Cookery

This new millennium promises extraordinary change in food. We will probably witness the global death of domestic cookery, as unheard-of foods fashioned from fungi and a host of other new sources come to the fore. Old-fashioned food will become a special treat. Today's junk food will disappear, as nutrient-rich innovations bring us a diet high in vitamins, minerals and health-giving supplements. The consumption of red meat (I write as an eater of meat) will rapidly decline.

Busy people in the modern world tend to choose a meal because it's quick and easy, no matter how much fat or salt it may contain. In future, food will be selected because of its power to boost the brain, or to delay the onset of senility. In the past, we ate in order to postpone starvation and gratify ourselves. In the future, we will choose food to enable us to relish life to the fullest extent. The capacity of future foods to enhance our lives and boost our bodies will make today's controversies over drug-enhanced athletes look tame by comparison. Everyone will be at it. Snacks designed to boost your mood, rather than to assuage your appetite, will be everywhere on sale. Controls on drugs have traditionally been very strict; but regulations governing foods and their additives are extremely liberal. Foods to boost your physique, your intelligence or your psyche are already taking advantage of this legal loophole.

We will need to introduce proper training in the future, of course. Those who sell drugs need special education, and qualified pharmacists are the only people who can supervise the dispensing of the most powerful medicines of all. By contrast, no training is needed for the people who sell products designated as health-food supplements rather than medicines. Because untrained people are cheaper to employ than those with degrees, health-food outlets are usually staffed by people who have little knowledge of what they are selling. The sale of some of these products demands careful counselling, for certain foods can react with a medication that a patient is already taking. Sometimes the symptoms for which relief is sought suggest a very serious medical condition which a doctor (or a pharmacist) would be able to diagnose.

Not only do we need new levels of education, but the commercial environment in which food is provided must be revolutionized. Although most red-meat producers have received steadily falling returns for their produce, the price paid by people in supermarkets has remained much the same. The suicidal farmers can be set in stark contrast to the wealthy shareholders who pocket the difference. In a culture supposedly founded on free enterprise, British farmers know that their prices are actually determined by a centralized bureaucracy based in Brussels. The supermarkets' profit margins are already higher in Britain than in the rest of Europe, so – as the farmers get poorer – the bureaucrats and the supermarkets are getting wealthier by the month.

Our present policies are also ruinous to health. If there are any components of food that most people believe to be harmful, then fats and salt would be among them. No matter: every lunch pack in the Western world contains a pack of potato chips or crispy snack-foods heavily dosed with both. These foods are potentially dangerous, yet nobody thinks twice before handing them out to their loved ones. Today's conventional snacks are puffed-up particles of industrial starch and chemical flavours, so rich in salt they could harm a child. In 1999 a case was reported of a baby who actually died through eating foods containing too much salt. Within twenty years, we will look back at them as we look back at adulterated foods from a century ago, when flour was expanded with liberal additions of sawdust, lead oxide and powdered chalk.

Salt levels can, of course, be reduced by using less salt in domestic meal preparation, but also by substituting potassium chloride for sodium chloride. It tastes similar, and the intake of potassium helps maintain the balance of electrolytes in the body. Lo-Salt is a commercially available product which contains twice as much potassium as sodium. In the future, many salty foods will use mixtures of potassium chloride to redress the balance. Is there a downside? Potassium salts are mildly radioactive, so that a bag of garden fertilizer rich in potassium is not something against which you should go to sleep, but then so are Brazil nuts and so too is our environment.

In the future, foods will be processed to ensure that they contain specific nutrients, the emphasis changing from year to year as new ingredients attract fashionable attention. Fat-reduced synthetic foods will replace the meat on which the typical Westerner has been raised. New cookery will blend the food of many different cultures to offer unheard-of tastes (I am writing these notes after a fine dinner in Chicago, with a curious menu that blends Japanese and French traditions). Such cross-cultural fusion food is set to expand.

As Efficiency Increases

Over the past half-century, global food production has undergone a colossal increase in efficiency. Yields of cereals have doubled in recent decades. In consequence, parts of North America and Europe have huge surpluses of food, and Brussels pays European farmers to destroy whole fields of grain crops where they stand because of over-production. The dead landscape that results is spiked with the seared remains of healthy young plants killed off in their prime.

This increase in efficiency cannot be sustained indefinitely at present rates, and in some areas (like fishing) a drop in production is more likely than an increase. This is why I foresee a move towards strange-sounding new foods. World economies are forcing small farmers out of business throughout the industrialized world. Many traditional crops cannot be commercially grown, because they are not on a governmentally approved list. In Europe, many rare or traditional crops cannot legally be sold, so seeds and tubers are passed around among enthusiasts. In future, the small farmer will become popular once more as a desire to experience old-fashioned organic foods begins to re-emerge. A specialist market in traditional treats will run alongside the great new industries that will feed the population.

Yet there are great problems lying in wait for today's mass-production food factory. Over a single generation, we have witnessed the emergence of new diseases. The newspapers are full of them – Listeria, E. coli O157-H7 (or VTEC, as we will encounter it in this book) and Campylobacter – yet they were unknown to microbiologists a few decades ago. New food brings new avenues for infection, and microbes are great opportunists. Thousands of people every year are dying because of infections brought about by present-day food production, and the future needs to address this issue.

Tightening Food Regulations

We have had epidemics of mad cow disease, (bovine spongiform encephalopathy or BSE) and salmonellosis in poultry. Beef from the US, produced with the aid of growth-hormone implants, has long been unacceptable in Europe. When, on the other hand, reports were published in November 1999 that human sewage and industrial waste have been used to feed cattle in France, this caused a tremendous furore. This distasteful practice seemed to contrast fundamentally with the French refusal to import British beef after the BSE affair.

Feeding sewage to animals actually has an ancient lineage, and is in principle very natural. All excreta is eventually metabolized, often by forms of life that are very different from those that produced it. Sometimes deposits are left behind, like the guano of seabird colonies, used for years as an agricultural fertilizer, or the fossilized excreta of dinosaurs which are known to us now as coprolites. In the main, however, but in the main, the eating of another organisms' wastes is a crucial part of the cycle of nature. For centuries Europeans dumped waste in the duckpond, knowing that it would eventually be used by small organisms in the water and eventually appear in reusable form in the form of fish and fowl. Chinese communities used to install latrines at one end of a fishpond.

Dried human waste contains about 50 per cent protein and 10 per cent fat, so it has the makings of a valuable food. During the 1970s British studies showed that, although sewage might be a useful food supplement for farm animals in theory, the idea could not be carried forward on aesthetic grounds. People would not wish to eat animals fed on human waste. Not only that, but in the sewers the human excreta is mixed with industrial waste, so sewage often contains high levels of poisonous industrial metals like cadmium and lead.

There are greater dangers. Many parasitic worms whose primary host is humankind, use cattle as the secondary host in their life-cycle. Human waste fed to cattle could perpetuate these parasites, some of which are picked up by humans from poorly cooked beef. We already excrete potentially harmful strains of *E. coli,* and the chance that some of these might elude the processing to which sewage is subject makes it increasingly dangerous to feed sewage to beef animals.

Human health can be compromised by industrialized food production, and our future regulations will have to change. Currently, we tend to tolerate diseases and take action only when people start to complain. In the future, I believe we must adopt a precautionary stance, and put safety first. Our controls must be based on intolerance of disease. Most Western nations tolerate *Salmonella* in chickens and scrapie in sheep, but this must change. Sweden has eliminated *Salmonella* from their poultry stocks. Australia has zero-tolerance for scrapie. It can be done.

We cannot continue to use science in the future as we have always done in the past. Science has been largely focused on reductionism, the quest for increasingly tinier causes at the expense of a broader vision, and this must change. What we need now is an holistic attitude to science. Separate scientific specialities are a luxury we cannot afford. I am convinced that in the future we need to bring together the disparate findings across the disciplines and marry them into greater truths. Multidisciplinary science and a new sense of academic urgency is what we will need to address these problems.

Food brings us some of the greatest delights we experience, and faces us with many of the most terrible problems we know. If humans are to survive in a crowded world, then the future of food must be very different from its recent past.

Chapter 2

Healthy Eating

They often say that half the women in the Western world are on a diet. That's misleading, for everyone is on a diet: it's what we eat. In the future we will become conscious of the extent to which our health, and even our survival, are predicated upon an enjoyable and satisfying food supply. Everyone will be consuming a healthier diet, though most will not realize what is in their meals.

The range of weight-loss diets that are currently available is confusing. For many people, adhering to a fixed list of do's and don'ts is the safest way to take control of what they eat. It is important not to go to extremes, like the followers of the 'breatharian' cult, who claim that you can live on nothing but air. Their leader, Ellen Greve of Brisbane, was challenged to demonstrate her diet on a television show in October 1999. After four days of abstaining from consuming anything (under close supervision), she was close to complete kidney failure and death. Her disciples said that this was due to pollution, rather than lack of food. The real lesson is that dieting should be taken seriously, for the health of the individual remains the top priority.

Storing for the Future

The development of civilization has been marked by its food. Humanity's first major technological achievement was the preservation of food for times when supplies were inadequate, by salting, drying and storing away. Humans are not the first species to discover this principle, of course; squirrels, for instance, know how to store nuts for the winter (though they often forget where they have hidden them).

Humans soon found out that food does not readily spoil if it lacks water. Now we realize that this is because the microbes that cause food spoliation need water to grow, and any way of limiting their water supply restricts their activities. In some cases this has been by drying food (like jerky and currants). Modern techniques include freeze-drying, where food is frozen and the water is driven off from the still-solid food. Products in this category can be rapidly restored to a remarkably fresh appearance by the addition of water. In other techniques, although the food still contains water, the water contains a high proportion of some other substance that prevents it being used by microscopic organisms. In salted foods (like cod roe and kippers) the salt traps the water and makes it unavailable for organisms that might cause spoilage; in jams there is too much sugar, and in pickles too much vinegar. These foods are safe to eat long after opening.

Frozen food has greatly increased in popularity ever since Clarence Birdseye popularized frozen peas during the 1920s. Water frozen into solid ice is not available for organisms to grow, so food stays fresh. Unfortunately, freezing can make meat and fish tough, and is best restricted to already-cooked products. Cook-chill foods maintain prepared products at low temperatures after cooking, though some bacteria (like *Listeria*) can grow in the refrigerator.

The introduction of canned food during the Napoleonic era sealed freshly cooked food away from the air, and the contents (if sterile when the can was sealed) would remain in good condition until the can was opened. There were occasional problems when *Clostridium* bacteria spoiled the food. *Clostridium* spores can withstand boiling; the bacteria grow without oxygen; and they can produce some of the most virulent toxins known to science. This is a potentially lethal combination, but modern techniques help to safeguard us from those risks.

The problems packaging will face in the future are twofold: first, people tend to think that wrapped food (like sliced meat) will stay fresh longer simply because it is bought in a pack. Secondly, the packaging plastics themselves can pose a medical threat. Modern Western men are producing less sperm than formerly. Under the microscope, human testicular tissue often looks unhealthy. It is now claimed that one of the causes is the abundance of phthalates in our modern environment. These are oestrogen-like compounds used as plasticizers in the manufacture of food wrapping.

Experiments with rats have shown that if one part in a million of these compounds is added to their drinking water, the rats' sperm production decreases by 15 per cent. As well as being used to make film wrap for food, a related compound, bisphenol A, is used to line cans of foodstuffs. Oestrogen-like compounds can leach from these cans into the food they contain. These are very potent chemicals, and they are increasingly widespread in the packaging industry. Future techniques must be developed to limit the chances of contamination from such packaging.

Fat Going to Waist

Green plants manufacture more complex carbohydrates and oils from sunlight than they require, many of which form a vital food reserve for periods when light or water may be in short supply. Within our own bodies the same mechanism is at work. Although we bewail the spreading waistline that so often accompanies middle-age, I believe that there is a biological purpose behind this extra layer of fatty tissue. The laying down of reserves of fat may be of crucial importance in sustaining our internal energy supplies during a wasting disease or a phase of famine in later life.

Our ability to manage our food supply has been a key factor in the colonization of the world by human beings, and the steadily improved diet of human society has been reflected in the rising levels of health and a consequent increase in longevity. Science has been behind much of this. As our understanding has advanced, we have come to recognize the crucial components of our food that are essential for health (like vitamins) and come to understand those (like saturated fats) that, in excess, can be bad for us. In recent years we have begun to create entirely novel compounds, like the synthetic fats that the body cannot use as a source of energy, but have also introduced into our diets traces of potent compounds (like the phthalates) that could have adverse long-term consequences on human health.

Food faces us with potent forces for good and bad, and in the future we will have to be able to balance these options in the interests of human safety.

Malnutrition in the Midst of Plenty

In recent years, Western medicine has experienced an unprecedented reversal in the steady progress of previous centuries. Diseases from our history texts, like rickets and scurvy, have begun to reappear in the West. In countries where food is available in abundance, malnutrition can be widespread. People eat fad diets, or fashionable foods, and although they take in more food than they need, they are losing out on vital nutrients.

Yet these deficient diets are not found where you might expect to find them. Vegetarians are rarely the victim of a deficient diet. People who become vegetarians simply by omitting meat from a previously mixed diet can have problems, but experienced vegetarians know how to ensure they obtain a healthy balance of essential foodstuffs. Their most likely deficiencies are of the vitamins B_{12} and D, which are abundant in meat, but surveys suggest that fewer than one vegetarian in a hundred suffers from a deficient diet.

Much of this is due to the progress of food science. Synthetic flavours and supplemental vitamins ensure that vegetarian products are nutritious and appetizing, and the vegetarian section in a modern supermarket is a tribute to modern technology. Products that look like ham, bacon or turkey (but are entirely non-animal in nature) are increasing in popularity, and are among the most 'unnatural' foods of all.

There are other sectors of society that encounter deficiencies. At least a quarter of tobacco smokers don't absorb sufficient vitamins B_6, C, E, or the folates (like folic acid, vitamin B_{12}). One-fifth of alcohol users share similar deficiencies, but are also short of vitamins A and B_1, and so are one-fifth of people on a weight-reducing diet, who are at risk from deficiencies of a wide range of vital nutrients. The contraceptive pill is associated with deficiencies of vitamin B_6, the folates and carotene (vitamin A) in 10 per cent of users, while a tenth of elderly people are seriously short of vitamins C, D and the folates. Diabetics are liable to deficiencies, and 4 per cent of pregnant women are also at risk. These are remarkable facts that we need to tackle in the future. It is now possible to produce appetizing and inexpensive foods that would rectify these problems. Above all, it is salutary to realize that tens of millions of overfed people in the wealthy Western world are still suffering from malnutrition as a result of their unbalanced diet.

Diet and the Elderly

To the elderly, diet becomes a matter of survival. There are some fortified milk-based drinks that elderly people and convalescent patients are encouraged to consume, but they often serve merely to rectify some of the deficiencies in intake and bring it nearer to the requirements for middle age. Older people will have to be aware of their need for specific nutrients if they are to enjoy the highest quality of life.

Surprisingly, some recent research suggests that advancing age encourages people to eat a broader range of foods. Vegetables become more pleasing to the tongue, researchers say, because sensitivity to bitterness in food seems to diminish with age. Women are notably less sensitive to bitter-tasting food as they become older, showing an increasing preference for vegetables, whole-grain foods, sour fruits like grapefruit and lemons, and beverages such as coffee and tea. Vegetables popular with adults, such as broccoli, Brussels sprouts and mustard greens, which can taste bitter (particularly when overcooked) can seem especially so to younger people, but their appeal to the elderly means that many older people often eat a healthier range of foods than they did when they were young.

However, there are also specific causes of malnutrition in older people. Disease may dampen down the appetite, and a diminution in the senses of taste or smell may make food less appealing. A loss of personal empowerment, like bereavement or isolation, can induce depression and a lack of interest in caring for oneself. Older people often also have less mobility and tend to stay indoors, and the resulting lowered exposure to sunlight can cause a lack of vitamin D. They may lack the energy to cook and prepare meals, to go shopping, or be unable to afford their favourite foods.

In many old people, poor teeth have militated against eating. In the West, the use of dentures is related to the year of birth, and people over

70 may have full sets of false teeth, whilst those born later have caps and crowns. Dentures are not renewed as often as is necessary – people neglect the fact that the mandible and maxilla are subject to bone loss and physical shrinkage as the years go by. Where they have been fitted, new dentures should be fitted at regular intervals, to take account of the way the jaws have shrunk.

Food intake is often lower in older people who are institutionalized. Studies in Holland and France have shown that levels of malnutrition increase once people go into homes. Partly this is because residential establishments frequently aim at unrealistic economies (thereby maximizing their profits) by underfeeding their older patients. One home for the elderly was said to feed all the uncommunicative inmates with bread and jam, knowing there was no way that relatives would ever find out. The lack of a stimulating environment for older people, and the loss of contact with their familiar way of life, can cause chronic misery, which results in a refusal to eat properly. People without personal power are more likely to feel ready for death. Refusing to eat is their only way to regain control over their fate. The valuable mental and physical resources locked away in old peoples' homes is an intolerable waste of human spirit and great wisdom.

A French study of 70 inmates of an old peoples' home showed that a quarter of them had clinical signs of malnutrition. A separate study of 204 old people in homes in France revealed that 27 were severely malnourished, and 40 per cent of the sample were showing some signs of dietary inadequacy. Italian nutritionists have found that more than half of the old people in their samples showed signs of deficiency. Future food will rectify deficient diets quickly and simply.

Enjoying Old Age

Very old people are often among the fittest individuals. It has been popular to believe that the diseases of old age steadily increase with advancing years, but research in America throws doubt on this convenient belief. Some studies have suggested that, whereas half of the elderly Americans in their eighties are suffering from Alzheimer's disease, only a quarter of those aged over a hundred show signs of the condition. When it comes to tests of cognitive, memory or manipulative skills, people around 100 score higher than comparable groups aged around 80. Although this seems paradoxical and counter-intuitive, the reason is that the younger age-group is weighted with a proportion of people with lower vitality, and they reduce the mean performance of the sample. Once you get past 90, you have shown yourself to be capable of successfully surviving such conditions, and the vitality which made this possible can stand you in good stead in later years.

Food Quality

A mere increase in dietary intake is never enough to overcome deficiency. Indeed, eating large meals can be a health hazard. Rats placed on to a semi-starvation diet live longer than those given food *ad libitum*. Tribal peoples whose diets are sometimes below apparent subsistence level (in European or American terms) can live long, energetic and healthy lives. The problem with deficient diets is not that people are eating too little food, rather that what they do eat is of the wrong kind.

In many cases, we are still discovering new minerals that influence our health. A generation ago, zinc deficiency was unknown. Since the importance of zinc was recognized, stunted growth due to a lack of zinc in the diet has been recognized from the Middle East and even in North America. Zinc is found in two forms of food: meat and wheat germ, the embryo portion of grain. In many developing countries meat is too expensive for all but the most wealthy individuals, and the embryonic portion of grain is polished off in processing.

Beriberi is caused by a lack of B vitamins, and is found wherever polished rice is a staple item of diet. There are plenty of B vitamins like thiamin in the whole rice grain, but they are lost in rice that has been polished. In some areas, rice is fortified with thiamin, iron and vitamin B_6, nicotinic acid. The supplements are sprayed on to the surface of the rice grains. Early attempts to fortify rice included riboflavin along with the other constituents. However, riboflavin is bright yellow, so the treated rice grains could clearly be seen in the final mix. The Indonesian women who were provided with this rice spent hours picking out the yellow grains and discarding them, so riboflavin is no longer included in the fortifying process. At the Swiss Institute for Technology in Zurich, geneticists have engineered a form of rice that produces vitamin A. A lack of this vitamin causes blindness throughout rice-eating areas, and this golden-grained rice could solve a pressing problem. The technology is being offered without patent restrictions, so that the benefits can reach as many people as possible.

Natural Food? What Natural Food?

Future foods will rely on fortification. The essential oils, vitamins, minerals and the rest will be readily available so that an impoverished diet is unlikely. There is one strident objection to this: that fortified and processed foods are unnatural, and unnatural foods are bad for us. I have much sympathy with those who warn against precipitate novelty in the food industry. In many cases we rush ahead, mindful of some current fashion, and the question of the health of the consumer is lost in the process.

The notion that food should be 'natural' is, however, unsustainable. Very little of our food could be claimed as natural. Being civilized is itself a highly unnatural state, and requires unnatural provisions to be sustained. A person who wished to consume a 'natural' diet should hunt wild animals, kill them with stones, tear them apart and make sure that the contents of the gut were consumed raw as well as the fresh, warm muscles torn from the carcass.

Many natural foods are harmful. A compound named phytate found in normal rice binds to iron in the diet, which leads to anaemia. It also hinders the absorption of zinc and copper. People who eat vegetables containing large amounts of phytate are at risk from this effect. Many vegetables (notably brassicas, like cabbage) contain potentially cancer-causing chemicals.

There is no strong case to make out for the supposed 'naturalness' of vegetarianism. People are biologically destined to be naturally omnivorous, as a brief consideration of human anatomy makes clear:

- Herbivores have eyes on the side of the head (to avoid predation); carnivores have eyes on the front (to catch prey with binocular vision). Human eyes are on the front.
- Herbivores have a large root canal in the teeth, which grow continually throughout life. Carnivores have no such canal at all; humans have a very small one.
- Herbivores have a large appendix. The appendix is absent in carnivores. Humans have a vestigial appendix which is so poorly developed that it is prone to infection.

And so on. The more we consider the anatomy and physiology of the human species, the clearer our omnivorous legacy becomes. Vegetarianism is potentially a liberating way of life if it eliminates the inhumanity of slaughter, removes from the diet a potent source of saturated fats, and reduces the calorific value of a diet. However, it is not somehow more 'natural' than the omnivorous habits of the global population.

Future food scientists will have to realize that abstract notions of a 'natural diet' are far removed from reality. What would you select as a traditional, natural meal? Wholemeal bread, butter and farmhouse cheese, for example? All are the products of technology which, even if it is traditional, is no less technological for that. They are completely unnatural, for nothing like them exists in nature. The grinding of grain to flour and its fermentation with yeast, prior to baking in an oven, are exceedingly unnatural processes. Isolating the fat globules of milk and patting them into butter is a technological procedure. Margarine, preferred by many, is a completely synthetic product, as well as being just as rich in fat and calories as butter itself. The spreads that are lower in calories are simply loaded with water (and cost more).

Cheese is a high-tech product. It is the result of lengthy processing of stale milk, as is the popular range of yoghourts now much in vogue. The setting of cheese into curd is through the use of rennin, which was traditionally obtained from calves' stomachs, but is now made by biotechnology. It may be that primitive tribespeople used to carry milk in pouches made from an animal's stomach, and found that it had the effect of curdling the milk (only a tiny trace of rennin is necessary to bring this about).

All cheese is made from milk from genetically selected cattle, which is kept until it goes off. In many cases, it is allowed to go off to such an extent that it becomes mouldy. In any event, milk itself is a profoundly unnatural food for humans. Unless you are a calf, cow's milk has no natural place in your diet. A significant proportion of the population is sensitive to milk; our obsession with this unnatural substance causes widespread ill-health.

Much of the problem arises from the fact that some people lack lactase, the enzyme which digests lactose (milk sugar). In Caucasians, about 10 per cent are lactase-deficient, though that percentage increases with age. Across Africa and particularly in China and Japan, the proportion is much higher. This is related to the fact that Eastern races regard the consumption of milk and cheese as unacceptable social behaviour. The majority of the world's adult human population is lactose-intolerant. This is hardly surprising, bearing in mind that no other adult animal naturally drinks milk.

Those who wish to make sure that food is nourishing and whole-some, free from harmful additives, and produced by tried-and-tested traditions have all my support. I am keen on such foods myself, and have travelled far and wide to bring home examples for my family to enjoy. However, naturalness doesn't come into it. Living in a cave, in the sleet and rain, without a supply of food is natural living. Subsisting on yoghourt and cheese, or bread and butter, is all sophisticated stuff; but it is no more 'natural' than making beer in a plastic pail.

Unnatural Farming

Few of our crop plants are natural. Wheat and rye, barley and oats are unnatural strains of grasses. Bananas and maize have been modified by tribal people for so long that these plants now rely on human inter-vention to propagate. The same is true of domesticated animals: sheep and goats, like cows and horses, are a far cry from the wild forms from which they were developed. Poultry, which many people regard as an acceptable form of meat – and as a half-way-house between an omnivorous diet and vegetarianism – is equally unnatural. The wild chicken is a woodland species from India. The instincts of hens, therefore, are to seek cover and the comfort of their fellows. A bare field, open to the sky, is a most unsuitable environment in which to raise contented chickens.

We often say that intensive farming ruins the land, but in my view any form of farming does so. When we grow a crop, we harvest it by removing it from the field in which it grew. Often, we eat only a small volume of the whole crop. For example, it is only the small grains of wheat we require, but we remove almost all of the plant from the field. There is a trend to dig the stubble back into the ground, but this takes time to degrade and the net result is a loss from the soil of nutrient with every harvest. Only sugar-beet is completely utilized, and even then the composted remains are not returned to the ground where they originated.

Farming is leading to the desertification of large areas of the planet. When the plants' food comes from fertilizer factories, the microbial communities that make soil into a living environment and the great web of wildlife are ignored. This is no way to sustain the soil for future generations.

Fish in the Future

Fish is an excellent food. Over a billion people in Asia rely on seafood as their prime source of animal protein. Not only is fish a good source of protein and vitamins, but oily fish are rich in compounds which we now believe to be valuable in protecting us from heart disease. The oily fish, like mackerel and herring, are not generally very popular. Tuna, a very fashionable fish in Western diets, is widely described in diet books as an oily species, but this is incorrect: the reason tuna has acquired this reputation is because it is commonly canned in vegetable oil, but tuna is not a rich source of fish oils.

The role of fish in our diet will have to change, largely because the oceans are being devastated by over-fishing. Fish farming is now providing plentiful supplies of cheap salmon and trout, and in the future we will rely increasingly on artificially farmed fish. We will also start harvesting alternative types of seafood, krill for instance, in order to supplement our food.

Reports from the United Nations Food and Agriculture Organization (FAO) conclude that 70 per cent of the world's fisheries are overexploited, and some have collapsed. At present we take about 90 million tonnes of fish from the world's oceans, and 70 per cent of this is concentrated in the hands of the European Union plus just nine other countries (in descending order: China, Peru, Chile, EU, Japan, US, Russia, Indonesia, Thailand and Norway). Over a third of what we catch is thrown back into the sea as unwanted. Another third is processed to make animal feed.

Factory ships have changed fishing dramatically. Gigantic trawlers remove huge catches from the seas. These days, half the annual catch around the world is caught by just one per cent of the vessels. Do not be tempted to imagine that the small, traditional boats are therefore innocent of causing unnecessary damage: the Italian driftnet fishing fleet in the Mediterranean Sea consists of small boats, which cast thousands of kilometres of drift nets every night. They are intended to catch swordfish, but they kill many other types of marine life, including sperm whales, dolphins and tuna. For many years the use of drift nets on the high seas has been banned altogether.

We have begun to see how important is this overfishing of the oceans. When the explorer John Cabot described Newfoundland in 1497, he wrote that cod were so plentiful that they almost blocked the passage of his ship. During the first half of the twentieth century, a quarter of a million tonnes of cod were taken from the abundant stocks each year. That ended in the 1950s, when the new factory trawlers arrived. They worked round the clock, in all weathers, scooping fish from the Newfoundland seas. The annual catch more than doubled, peaking at 800,000 tonnes in 1968.

Clearly, this was not sustainable. By 1975 the catch had fallen back to 300,000 tonnes and during the following year the Canadian authorities increased their national jurisdiction to 200 nautical miles. Canadian ships, using huge dragnets, continued to fish their coastal waters until by 1992 the government was forced to end the fishery altogether. Two years later it was estimated that the cod population amounted to no more than 1700 tonnes. Thirty thousand people lost their jobs after the 1992 cod moratorium took effect, and they were followed by an additional 12,000 fishermen and plant workers after these additional cutbacks and closures. Newfoundland had lost 40,000 jobs and an entire industry was brought to a halt.

If the authorities had acted in the 1970s, it is more than likely that the future of the fishery could have been assured. Now we see similar signs elsewhere around the world, and the future of our fisheries depends upon conservation. A large fishery based on the Georges Bank off Massachusetts was the next to close. Off the British North Sea coast, cod, hake, herring, haddock, mackerel, plaice, saithe and whiting are all now at dangerously low levels. There are over 200 species of fish in British waters. Seals, porpoises and dolphins, as well as giants such as orca, minke and fin whales are found in the seas around Britain, all of them supported by the fish stocks. London apprentices used to complain of their diet of salmon, and oysters were the food of the poor. There was a British tuna fishery until the 1960s, and the world record Blue Fin Tuna was caught off the Dogger Bank in 1938. Tuna are now absent from the North Sea.

The future of our fish supply depends upon management. We can increase fish farming, but we crucially need to establish sanctuaries – 'national parks' – in the oceans where fish are free to breed and multiply. Large areas need to be left to recuperate, and in future we must ensure that catches are no more than the fish stocks can tolerate. Fish are crucial players in marine ecology, and too important a component of our diet to be abused any longer.

Hidden Dangers

Today's diet is tainted. Traces of potentially harmful chemicals are widespread in our food. The dioxins include some of the most toxic by-products ever created by the chemical industry. The most potent is 2,3,7,8-tetrachlorodibenzo-p-dioxin (or TCDD) and this is what we generally call dioxin. It was present in Agent Orange, used as a defoliant in the Vietnam War, and is believed to have caused many of the medical problems suffered by veterans. Traces of dioxin have been found in tampons, and a range of conditions due to dioxin have been found in animals, including foetal death and deformity. It is now believed that dioxin causes cancer in humans, and it may even be linked to conditions like endometriosis.

In May 1999 warnings were issued that huge amounts of dioxin-contaminated animal feed had been sold throughout Europe, principally in Belgium. The sale of chickens, eggs and their by-products was immediately curtailed, and two Belgian cabinet minis-ters, who were said to have known about the problem for weeks without telling anyone, were forced to resign. Within days, Russia banned imports from the EU, and in June many nations in the Far East, including Thailand, Malaysia, China and the Philippines, imposed bans of their own. At one stage, supermarkets in Thailand even removed Belgian fly-spray from their shelves.

Millions of people will have eaten food contaminated with small amounts of dioxin. We will never be able to decide whether a specific case of illness was linked to the contamination of the patient's diet, but with such potent chemicals we need to take care. Huge problems would result unless food scientists of the future ensure that their products are safe and free from the cancer-causing by-products of this industrial age.

Additives and the Future

The term 'chemical additives' is meaningless. Humans are bundles of chemicals ranging from potassium and nitrogen to sodium, calcium, carbon, oxygen and the rest; we are chemical systems above all else. From the point of view of our chemical constituents, you and I might as well be an outcrop on the far side of the moon.

Food is also composed of chemicals. As for additives, well, the sodium chloride artificially added to pretzels or fish-and-chips sounds intimidating until we recognize it as table salt. The problem with the chemical industry is that it needs to sell its products to make money. In consequence, huge numbers of superfluous chemicals are regularly supplied to us in our diets, and not all of them have been thoroughly tested for untoward effects. Hyperactivity in Western children is claimed to be on the increase, and it has been correlated by some authorities with unnecessary food additives. Tartrazine, for example, has been claimed to be a cause of problems in the classroom, and children who transfer to a diet free from such dyes have been reported to be calmer and more rational. These childhood effects may have consequences in adult life.

Future nutritionists will need to resolve the confusion on safety. Sales of vitamin B_6 were restricted in Britain in 1999, but soon came back on the 'favoured' list. Similarly, beef on the bone was removed from sale to the British public on the basis of scientific evidence, while European Union scientists declared that British beef was 'safe'.

Americans rely on cyclamates as artificial sweeteners; they cannot be sold in the UK, although the evidence against them was never satisfactorily confirmed. Conversely, saccharine – popular in Europe – has been associated with cancer risks in North America. Most nations rely on aspartame as a non-calorific sweetener, and it is widely regarded as inherently safe because (although it is tasted by the tongue) it is said not to enter the bloodstream. Recent research in Barcelona suggests that aspartame may be related to increased levels of brain cancer in rats. Could it apply to humans? Critics of the reports claim the research is flawed, and state that the molecule is inert in the body. However, if a substance is sufficiently soluble to occur in drinks, and to be detected in its dissolved state by the taste-buds, there might be mechanisms by which it could interact more intimately with living cells. Some nations are not so easily convinced of the innocuous nature of these compounds, and the government of Iceland even imposed a ban on the sale of aspartame in the autumn of 1999.

The future of food depends on sound and solid scientific evidence. Being civilized is itself a highly unnatural state, and our survival in the centuries ahead will depend upon the best of science and the most honest of intent. With fanatical lobbying on one side, and a chemical industry hungry for profits on the other, it is easy for the public to be left out of the equation, and for truth to become obscured.

CHAPTER 3

FOOD AND CULTURE

There are very few university departments concerned with nutrition, and some of those have been badly served by their academic masters. As a result of a fashionable belief that nutritional science was not worth supporting, the Dunn Nutritional Unit at Cambridge (England) was peremptorily closed down in 1998. So drastic was the change in policy that one of their leading scientists says that he was told never to enter the premises again. A generation ago doctors were taught nothing on the subject. Even today they are taught remarkably little. Nutrition remains a Cinderella science.

Meanwhile, we are bedevilled by warnings about the dangers of what we should *not* eat. No health authority proposes that we eat a specific food because it is enjoyable; rather, they warn us *against* foods with dire warnings of death and disease. This is the age of the threat, and people in the West exist under a cloud of terrible warnings and frightening innuendo. Stress and tension are implicated in causing illness, and I believe that much modern health propaganda simply adds to that problem. Rather than reducing problems, these strident warnings about food safety add to the burden of human suffering.

Future Foods and Heart Disease

'Cholesterol causes heart attacks, so must be removed from our diets.' Few compounds have had such a bad press as cholesterol, yet this is a vital substance in our bodies. Cholesterol is an essential constituent of the insulation that coats our nerves. It is as vital as the plastic insulation in a telephone exchange. The breakdown of the insulation layer short-circuits nerve impulses and produces devastating consquences. In multiple sclerosis it is the cholesterol-rich insulating layer that disintegrates and causes the symptoms.

There is no mystery about what cholesterol actually is: its chemical formula is $C_{27}H_{45}OH$. The name of the molecule from which it is derived, if you prefer things in words, is cyclopentanoperhydrophenanthrene. It was first investigated in the late nineteenth century, and by 1932 its formula had been deduced. Cholesterol is a white, waxy solid with a soapy feel which is virtually insoluble in water. Although cholesterol comprises only two parts in a thousand of the human body, it is found in every living cell and is vital for survival. We know that it is synthesized in the body; one early experiment showed that mice fed on a cholesterol-free diet produced just as much cholesterol in the growing bodies as those fed cholesterol in their diet. Cholesterol is produced in your body at the average daily rate of 10mg per kilogram of body weight. If you eat a diet rich in cholesterol, the amount produced in the body goes down; if you eat too little the body produces more.

For this reason, diet exerts a relatively small effect on cholesterol levels. You produce far more of your own cholesterol than you eat. It is often claimed that high blood cholesterol levels promote atherosclerosis and consequent coronary heart disease. But many studies have shown that people whose blood cholesterol is normal have just as much atherosclerosis as people whose levels are raised. Medical surveys have revealed that people who have had a heart attack had not eaten more fat than other people, and that the extent of atherosclerosis at autopsy is related to many factors other than diet.

You might think that a cholesterol-lowering component of food would be important in the future. Since the only effective way to lower cholesterol is with drugs, known collectively as statins, surely future food would need to include products that contain these substances? Dr Uffe Ravnskov, a Swedish specialist, has surveyed the literature and concludes that neither heart mortality or total mortality have been improved with drugs whose effect is only to lower the cholesterol level. 'On the contrary,' he emphasizes, 'these drugs are dangerous to your health and may even shorten your life.'

Studies have been published of the incidence of heart attacks in many countries, but it is hard to make sense of the results. In Finland for instance, mortality from heart disease in one surveyed area was five times greater than in another, despite people eating much the same diet. In Greece one area showed seven times as much heart disease than another, although other risk factors were similar.

Benecol is a synthetic oil which is the first popular food to be associated with lowering the cholesterol levels in the bloodstream, and its approval in the United States was soon followed by release in Europe. Margarine made with Benecol is far more expensive than conventional brands; clearly there is a perceived premium in a food designed to regulate the *milieu intérieur*.

Meanwhile some recent research suggests that soya protein may help protect against heart attacks. Over fifty independent scientific studies have confirmed that as little as 25gm (about an ounce) of soya protein per day, as part of a low-fat diet, can reduce the incidence of cardiac disease. Soya milk is becoming increasingly popular, and with its low levels of saturated fats it has long been seen as a viable substitute for dairy milk. The value of soya protein in protecting against one of the greatest killers of the age will lead to a considerable increase in production. The United Soybean Board in the US are planning for as doubling of demand within five years.

Polyunsaturates

Polyunsaturates are popular. People believe that they protect against heart disease. It has been shown that, in some cases, an increase in the proportion of polyunsaturated fatty acids in a high-fat Western diet can apparently lower the levels of cholesterol in the blood. However, some people are unaffected. For many years health authorities promoted polyunsaturated fatty acids (PUFAs) as a vital health supplement, but later research has cast doubt on this conclusion. A high intake causes rats to develop cancer, with damage to the testes and infections. Modern recommendations are that no more than 7 per cent of energy should come from PUFAs. These views are still changing.

Rapeseed oil has become increasingly popular because it is high in unsaturated fatty acids. However, it also contains high levels of erucic acid. Because our usual enzymes cannot tackle this large molecule, its presence may cause problems for people with weak hearts. Clearly, the mere presence of unsaturated fatty acids is not always a clear-cut benefit.

One of the most popular PUFAs at present is gamma linoleic acid, GLA. It is an important compound in the production of other substances, like prostaglandin, which protects against high blood pressure, high cholesterol and the production of dangerous blood clots. It is produced in the body from linoleic acid, but the conversion process is inefficient. Many conditions, including a lack of vitamins or zinc, or a high consumption of alcohol, militate against the synthesis of this valuable molecule in the body. Low levels of GLA in the body have been linked to diabetes, dermatitis and even cancer.

Breast milk is rich in GLA, though it has not traditionally been incorporated in formula milk for bottle-fed infants, an omission that is overdue for rectification. Some continental European brands of baby milk formula now include GLA, and this may be a pointer to developments in the future. Once we understand the action of these compounds, food products of the future will contain substances like GLA to help our bodies to maintain levels of the crucial components for our metabolism and to resist disease.

Survival of the Fattest

Fat is not all bad news: in future we will understand how crucial it is for survival. Caroline M. Pond, a British nutritionist, even argues that it is a taste for fats which may have led to the development, in humans, of a sociable personality and quick-wittedness. She states that people with higher blood cholesterol have been shown to be capable of faster mental processing, and reports that slimmers can complain of tiredness, depression and irritability.

Future foods will need to incorporate vital forms of fat, and food commentators will need to be more balanced in making their pronouncements. The popular belief is that saturated fat is animal fat, but this is a misleading generalization. Palm and coconut oils are primarily saturated too, and palm oil is one of the most widely used of all plant oils. Meat-free diets based on these plant oils can still be high in saturated fats.

We also accept that saturated fats cause an increase in heart disease, but in England the amount of animal fat in the average diet has been fairly constant throughout the twentieth century, while the number of heart attacks increased tenfold between 1930 and 1970. There must be other factors at work. We find a similar pattern in the United States, where mortality from coronary heart disease also increased ten times in the period 1930–60 although the amount of animal fat in the diet stayed much the same. There was a steady decrease in deaths from coronary heart disease in Switzerland after the Second World War – but during the same period, the intake of animal fat went up by 20 per cent.

The craze for beef causes very severe pressures on the developing nations. Twelve million hectares (or 30 million acres) of rainforest are being destroyed every year, and 90 per cent of that is because of the spread of grassland for raising beef. In Guatemala, much of the land is now devoted to raising beef, 18 million kilograms (40 million pounds) of which is exported to the US each year. Yet three-quarters of the child population of Guatemala is undernourished. Five years ago, 6 per cent of the grain in Mexico was consumed by beef cattle; now the total is 50 per cent and rising.

A kilogram of grain will make a kilogram of bread, but it takes sixteen kilograms of grain to produce a kilogram of beef. Every time a person in the West decides to give up eating beef, 3000 square metres (35,000 square feet) of land are freed up for crop production. In real terms, meat prices have steadily fallen. In 1963 the retail price of chicken meat in the US was $1.12 per pound. By the end of the millennium, and adjusting for the rate of inflation, it was one-third of the price. Beef prices declined by about a quarter over the same period.

Little wonder beef remains a popular food. Now that too many Western people are overweight, it is clear that future food will contain less saturated fat than the present-day diet. We will also see the development of health-promoting foods that contain specific fatty acid molecules. It is important to realize that fatty acids are not present in

fresh vegetables, but result from the breakdown of natural components of the cells. In future we can ensure that our diet includes fats
found to be important, and limits others which could harm our health.

We can also start to take advantage of the compounds in the body
that regulate fat. The first of these is leptin, a messenger chemical that
regulates obesity. It was discovered in 1994, and was soon found to be
directly involved with the control of obesity in rats. A leading bioscience company paid $150 million for the rights to the discovery, only
to find out that leptin didn't work the same way in humans.

Even so, it does reveal something about the control of the malignant
form of obesity that sometimes causes children to become dangerously fat. These children lack leptin altogether because of damage to a
chromosome, and the rate of growth is life-threatening. A little girl,
normal when born, weighed 16 stone or 225 pounds (102kg) by the
time of her ninth birthday. Another patient weighed five-and-a-half
stone, 88 pounds (40kg) at the age of three. By their teens, these
children need liposuction before they can walk.

Their disorder gives them a ravenous appetite, and they are not
satisfied with consuming less than 2000 Calories at a single meal.
When they were treated with leptin, their appetite fell dramatically and
no more than 300 Calories were needed at a single sitting. The disorder delays sexual maturity, so that adult women in their twenties
have still to start menstruating. Leptin treatment restores sexual development to normal. Unfortunately, administering leptin has no effect
on normally overweight people, so the hope that we had a quick fix for
fatness has not been realized in practice. None the less, we are on the
way to recognizing the compounds that regulate fatness, and in the
future we will be able to treat people so that they can remain healthy
and avoid the complications that obesity can bring.

Many Western adults are overweight. Britain has just become the
fattest nation in Europe, beating Germany into second place late in
1999. The prevalence of obesity in the UK is the same as it was in the
US a decade earlier, and the costs are high. The British National
Health Service now pays £1.7 billion ($2.8 billion) treating obese
victims of cardiac disease, cancer and diabetes. The care bill for obese

people is larger than that for victims of tobacco smoking. Further-more, in the US at least one child in five is overweight and the number continues to grow. Over the last twenty years, the proportion of over-weight children has increased by more than 50 per cent, and the number of 'extremely' obese children has doubled.

The increase in obesity in the English-speaking world surely cannot be due to eating too much. People eat no more now than they did several decades ago. A more important factor is the lack of exercise. Youngsters who used to walk or ride bikes to school are customarily driven there, and picked up afterwards. Many sports activities are spurned by the young. The pressures of the world of the future will encourage people to collapse in front of a TV set rather than go walking or to the gym.

Stanford University has shown the effect that a change of life-style can exert. They studied third- and fourth-grade children at two elementary schools. At one school, children watched television and played video games as usual. At the other, the researchers initiated an intervention programme. This consisted of a series of 18 lessons on topics such as self-reporting of TV and video games, becoming 'intel-ligent viewers' and advocates for less TV time when shows did not deserve attention. During the 'television turnoff' phase, two-thirds of the kids said they watched no TV or played no video games for ten days. Parents were supplied with an electronic time manager that monitored each household member's viewing time.

The researchers found that the kids who participated in this inter-vention programme had significant decreases in obesity measures when compared to their couch-potato peers. It is important to empha-size that kids in the intervention group ate just as much high-fat food as those in the control group, and they did not take up any new physical training activities. The simple act of giving up a sedentary occupation brought about the improved the state of health.

The French Paradox

French people (for all their rich food) have much less heart disease than the British and Americans. A popular view claims that red wine is the answer, because it is full of healthy components which keep the arteries free of deposits and give a youthful zeal to everyone who drinks it. The latest theory has been published in the *British Medical Journal* – doctors now claim that the reason the French have so little heart disease isn't anything to do with drinking red wine, after all. The reason they give is because the French only recently started eating a lot of saturated fat, and their data is simply waiting to catch up with the British data. They say, it's not the wine which controls the levels of heart disease, it's the diet.

This thinking, in my view, typifies the oversimplified approach of much modern science. People can only handle one idea at a time, but life is multi-factored and many-faceted. Research is conducted as though there is only one factor behind heart disease. There must be many. I can list some here:

- We must consider the diet of the foetus – even nutrition in pregnancy may have a part to play.
- In India, heart disease is rapidly becoming the number one killer, in a nation where almost nobody eats beef.
- What of the role played by microbes? An organism named *Chlamydia pneumoniae* has recently been identified in many post-mortem specimens from victims of heart disease.
- One of the reasons why heart disease has decreased in some populations may be because antibiotics have been widely prescribed over the past decades, helping to wipe out these infections.

Let's try to make sense of all this. While everyone seems to plump for a single cause of the French Paradox, it is revealing to take stock of some of the other issues. Although French people drink lots of red wine, their overall life expectancy is comparable to those observed in Britain and North America. Their love of wine leads to raised levels of alcoholism, for one thing, which balances out the supposed benefits. As we have seen, *Chlamydia pneumoniae* bacteria have been found in plaques that form within the coronary blood-vessels of heart-attack victims. We need to know whether this species as widespread in

France as it is elsewhere. Does the French way of life hinder its transmission – is there something in the diet which acts against the germ, for example? Have antibiotics knocked out the organism? Or has the French way of life acted against the spread of *Chlamydia pneumoniae*? It may be that the rise in heart disease during the twentieth century was due to an epidemic of *C. pneumoniae* and not to diet after all.

Then there is the question of psychology. Several surveys have shown that there is a relationship between hostility or depression and the incidence of heart disease. Clearly, one's state of mind can also play a crucial role in the health of one's heart. As a family with homes both in France and Britain, we do see the differences. The French are, generally speaking, a confident lot – their detractors would describe them as arrogant. If depression is related to heart disease, then high levels of self-esteem might well have the converse effect.

We can also consider questions of lifestyle. The French have a far better developed social structure than many Western nations. When do people drink wine? When they are with friends, as a rule. If you invite people round, or spend time sitting discussing affairs of state or the problems of the young, then you have time to open good wine and drink it socially. It may be that a relaxed and cheerful attitude to socializing is the real protective against heart disease.

Starch Wars

Fibre has had an unenthusiastic press for decades. It has been claimed that fibre in the diet could help prevent major diseases, like heart attacks and cancer. This all seemed too far-fetched for the medical establishment, and the notion of fibre in the diet was widely regarded as a trifle eccentric – like an exaggerated fixation on bowel movements. It is now emerging that the important components include *soluble* fibre. Oats are a particularly rich source, which explains the value of porridge in the diet, though the Scots (for whom porridge is a national dish) suffer from high levels of heart disease said to be due to heavy smoking and alcohol consumption, and to an unhealthy diet.

Soluble fibre is composed of complex carbohydrates, starch molecules so large that they are not easily digested, but pass through to the large intestine. Polysaccharides in the soluble fibre are important in providing food for the health-promoting bacteria in the gut. I call these organisms *salugens,* from the Greek *salus* (health), to distinguish them from the familiar infection-causing *pathogens,* from the Greek *pathos* (disease). Salugens promote health, just as pathogens make us ill. In the gut are many such salugens, including the Y-shaped bifid bacteria which seem to protect against cancer. Eating soluble fibre helps to feed them, and in turn they protect us against disease.

Food as Culture

We have seen how food and human society are inextricably related, but there is a deeper aspect that the future will have to address – not food and culture, but food *as* culture. We must recognize that meals are not only for nutrition: while food is for feeding, meals are also a form of societal communion. Food is a great indicator of perceived social status, and innovations will have to weigh up these factors with care.

For decades black coffee became increasingly popular in the Western world, and partially eclipsed the traditional cup of tea in Britain. Except for frothy cappucino, coffee with milk was unfashionable, but when caffè latte appeared, how things changed. Milky coffee (dignified by its trendy name) was suddenly the height of social acceptability and, propelled by casual references on the cult American TV show *Frasier,* caffè latte made a desirable beverage out of an unsophisticated drink suitable for children.

There are now popular sayings that perpetuate and reflect fashionable food trends, like 'real men don't eat quiche'. This is a particularly fascinating example, and if we think it through we can perceive that the preconception belies the reality. There is one dish we have been asked to abjure: the traditional British fried breakfast of bacon or ham and eggs, rich in fat and starch. Health pundits insist that this is the definitive unhealthy meal.

Look at the menu of a health-food restaurant, and there is one item that is always present. It is the hallmark of modern healthy eating – quiche Lorraine. To begin with, it is not a 'quiche' at all. Should you visit Lorraine, that pleasant, pastoral *département* of northern France, you'll find that their open tart is light and refined, an example of French peasant cooking at its best. The UK quiche is nothing more than egg flan. We do not call it that, of course, because the word 'flan' connotes heavy, starchy, old-fashioned food, whereas 'quiche' sounds modern, foreign and fashionable.

Consider the ingredients. The pastry case is made with fat and flour. Then there's the filling: egg, a little milk, chopped bacon. It comes from the oven looking golden and delicious, and much more healthy than the old-fashioned fatty and starchy traditional breakfast of bacon, egg and fried bread or hash browns. Wait, though; didn't we encounter the same essential ingredients in the quiche? It is just that quiche is fashionable, and the bacon-and-eggs breakfast is not.

The future choice of names for dishes will be a matter to preoccupy tomorrow's caterers. Although we delude ourselves into believing that a traditional fried breakfast is inherently bad, whereas a quiche is fundamentally healthy, they are much the same. A key issue for the food of the future is what we choose to call it.

Refined Food, Refined Customers

People will always seek out speciality meals, delving back into the past, patronizing local producers and teasing the palate with special concoctions. However, the artificiality of civilization has made us into refined people and we will have an increasing need for refined food.

Are refined foods inherently undesirable? Why should popular opinion hold that brown bread or brown sugar are so much healthier than the refined, white versions? Pure white sugar, like pure flour, used to be in demand because, in previous centuries, such products were widely bulked out with impurities. Most contaminants (rust, droppings, wastes) are brown. Refinement was a move towards purity, and the evidence was visible in the pure coloration of the product. Food continued to be adulterated (alum was added to bulk out sugar just as chalk was put in flour), but sugar darkened with dirty water or animals' urine was a health hazard, and white sugar was clearly better. Similarly, today's wholemeal bread is speckled with added husks, bran, seeds and grain, not the impurities of old, such as crushed beetles. White processed bread, however, looks purer, and is the result of decades of detailed research into the quest for efficiency. White bread is now bulked out with water, which often comprises more than 45 per cent of the loaf. I was once told by a director of one of the bread-making companies that the quest was to add more water to the product; he added that 49 per cent was the most you could obtain before the loaf collapsed. 'Our holy grail is to reach the magic 50 per cent,' he said. I told him that this would not be wise. The modern convention, when listing ingredients, is to list them in order of their abundance in the product. Were a bread-making factory to break the 50 per cent barrier, then water would be the main ingredient. They would be selling 'sliced water (with added bread)'.

Soft white loaves are more to the American taste, but in Europe they are widely regarded as an inferior product. Similarly, the colour favoured for eggs is related to culture. Americans like to buy white-shelled eggs, for example, whereas in the UK shoppers prefer mottled brown, believing that this is a sign of 'naturalness'. People believe that the golden-yellow yolk is also a pointer to a healthy and natural egg, but they are wrong. Many industrially farmed hens are fed a diet containing canthaxanthin (E 161g) which gives the yolk a golden-yellow hue. Questions have been raised over its safety, so this compound is being withdrawn, but the caroteinoids that are added to hen-food provide virtually any coloration you like. Egg producers can choose from a colour-chart, matching colour to the feed the buy for their hens. Natural eggs actually have pale yolks. In future, when the facts are more widely known, purchasers will come to regard a pale (rather than a golden) yolk as the indicator of a naturally healthy egg.

Food Culture in the Future

In the twentieth century, tables were laid for meals with formality. Knives and forks were the hallmark of civilized eating throughout the West, but finger food (popular in many other parts of the world) is spreading widely in Western cultures. It is a young and fashionable way to eat, and is also a lot cheaper for the caterer. The influx of Eastern cultures has already affected Western eating habits. In the future we will see more chopsticks in the West, and more Western conventions in the East.

Japanese cookery will become far more popular: it is highly decorative, with as many as fifty dishes for a formal reception. As we shall see in Chapter 8, foods that seemed strange and alien in past generations will become readily accepted in the future. Some are already close to home. There are ancient foods that seem inherently repugnant, such as rotten fish. Could anything seem more alien to the Western palate? Surprisingly, such foods are highly prized around the world, and do have a legacy in the European diet. The Laps eat rancid fish, putrid and odiferous but highly valued. Ancient Rome served a sauce of rotten fish with every savoury meal, and brought the convention to Britain. It survives in Worcestershire sauce, which is made to this day with fermented fish.

The future will need to recognize a false orthodoxy: the rule that children eat what their parents ate. Were this true, British schools would not have menus featuring lasagne, curries, pizza and baguettes. American school students would still be eating fish balls, rather than Taco Bell. The hamburger would not exist, and neither would the sour sweets and bubble-gum so popular with every American youngster. In truth, people eat what they believe will benefit them – not necessarily by improving their health, but by asserting their social identity. The future provision of a healthy diet will be based upon a full and detailed understanding of these motives.

CHAPTER 4

FOOD DISEASES OF THE FUTURE

Our safety and survival in the future depend upon food free from infection. Large and centralized food production embodies the potential to spread disease to tens of thousands of people in a matter of hours. Our understanding of food-borne diseases remains incomplete. Meanwhile, the problems are increasing. New diseases spread by food are appearing all the time.

This comes home to me in a practical fashion. As a student I was dissatisfied with the level of understanding of the way microorganisms relate to our food – how microbes help us to make food, how it is produced, how it feeds us and yet how microbes can make us severely ill. It seemed to me that the best way to tackle this problem was to write a new type of microbiology textbook, and my book *Microbiology and Food* (1970) was the result. It was received with enthusiasm, and since then I have often been asked about republishing it. It wouldn't be worthwhile. The book may have served a purpose in its time, but it is pretty useless for today's student of food science. So many new infections are being documented that my list of diseases would be incomplete. It is popular to believe that the main food diseases were documented long ago (in the nineteenth century, like most other bacterial infections) but for food-borne illnesses this is untrue. As we move into a well-fed and highly productive era, new diseases transmitted through our food are regularly emerging.

Incomplete Immunity

Curiously, some of our problems result from the way in which civilization has removed us from contact with the familiar organisms our predecessors encountered in their daily lives. We do not develop childhood immunity to many of the microbes which we may encounter.

Perhaps problems with immunity account for diseases of the intestinal lining like colitis and Crohn's disease. In earlier centuries, tapeworms and threadworms were regularly found in the intestine. Tapeworms were even used as a slimming aid prior to the First World War. Tablets offering guaranteed weight-loss were sold, each containing a tapeworm's head. The worm would establish itself in the gut, competing for food with its host, who soon lost weight. Strong medicines were taken afterwards (hopefully to get rid of the worm) leaving the hapless host a lot thinner.

Our immune systems used to encounter intestinal parasites as part of normal development. Modern society has largely eliminated worms in the West, so the immune system is no longer used to these gut parasites. The lack of contact with the worms with which we evolved for two million years has unbalanced our immunity, so one theory goes, and one experiment has even shown that parasitic worms introduced into the intestine can induce a remission in disease.

Another example is the water-borne pathogen *Cryptosporidium*, a minute organism found in the excreta of farm animals. In earlier times, it would have been encountered early in life. Its effects are a week or more of diarrhoea (which is the bad news) followed by lifelong immunity to the organism (which is good). These days, it still runs riot in farm animals, but usually most people do not encounter it at all. In modern societies, people are kept apart from *Cryptosporidium*.

Sometimes flooding or an agricultural accident leads to animal waste being washed out from a farm into a nearby watercourse. On occasion, the contamination spreads to food or (more likely) to the water supply. *Cryptosporidium* is notoriously resistant to conventional methods of water purification. The organism now threatens older people, who lack the immunity found in traditional communities. Its effects are far worse. Among the elderly, or those with a compromised immune system (like victims of AIDS), it can kill.

New Diseases

The proliferation of new diseases is disturbing. In the future, as we face more diseases from food than we did in the past, we will be witnessing a reversal in a thousand years of medical history. Tomorrow's food-borne diseases will include these emergent problems:

- *Listeria* – a serious 'flu-like disease which kills a third of its victims.
- *Campylobacter* – a recent discovery, caught from chickens, etc.
- VTEC (certain strains of *E. coli*).
- BSE – believed to have caused $_{nv}$CJD (new-variant Creutzfeldt-Jakob Disease).
- *Salmonella* – 1 hen's egg in 1000 carries this serious disease organism.
- MRSA (methicillin-resistant *Staphylococcus aureus*) – can grow in whipped cream, generally incurable.
- *Mycobacterium* – the genus that includes the TB germ, now re-emerging in the West.
- *Clostridium perfringens* – a very common bacterium that can sometimes cause diarrhoea and severe cramps.
- *Bacillus cereus* – causes mild illness, is now often found in baby formula milk.
- *Vibrio parahaemolyticus* – contracted from coastal fish and shellfish; causes gastroenteritis.

We can see how rapidly matters have deteriorated if we look back half a century. The only bacteria known to be spread by food in 1950 were *Clostridium botulinum*, *Staphylococcus aureus* and *Salmonella*. The first two are abundant in nature and are normally harmless.

Clostridium botulinum occurs in the soil and it is the cause of botulism. *Clostridium* bacteria can grow only in the complete absence of oxygen, so they have been found in contaminated cans of food. The toxin they produce, botulin, is one of the most toxic compounds known to science.

Staphylococcus aureus is familiar as the bacterium that causes boils. It lives on the skin, and usually does no harm. However, it can grow in cream at room temperatures and produces a toxin as it multiplies. People eating contaminated whipped cream quickly become ill with stomach pains. Although penicillin became best known for its ability to cure infections due to *S. aureus*, some strains of these bacteria are now resistant to every normal antibiotic. They are known as MRSA for short ('methicillin-resistant *Staphylococcus aureus*'), and are rapidly spreading in hospitals around the world.

Salmonella is not normally found in the environment, but many of the species of this important genus of bacteria cause short-lived infections in the intestines with diarrhoea and pain. This is the organism that often infects poultry farms. *S. typhi* is the species that causes typhoid fever, though there are large numbers of different strains presenting us with differing degrees of danger.

By the 1960s, *Clostridium perfringens*, *Bacillus cereus* and *Vibrio parahaemolyticus* had been added to the list of bacteria spread through food. More recently, *Listeria*, *Campylobacter* and *E. coli* O157 have been recognized. Others are likely to be added. As future food products become diversified, so will the means for infection.

The bacterium *Listeria monocytogenes* is not a new discovery, for it was first identified in 1926, but its role in causing food-borne infections is far more recent. In 1985 a supply of contaminated Mexican-type cheese led to about 150 cases of listeriosis in the United States. There were at least 48 deaths. Soon afterwards there was an outbreak of 122 cases with 34 deaths in Switzerland, which was spread by Vacherin Mont d'Or cheese. In France in 1995 Brie de Meaux cheese led to an outbreak with 17 victims; it killed one old person and led to four aborted pregnancies among the nine pregnant victims in the episode. Not only does *L. monocytogenes* kill many of its victims, but it has an incubation period of up to 70 days. This makes tracing the origin a difficult matter. Although the bacteria can multiply in soft cheeses, they also can occur in hard cheese without posing such a danger. Health authorities allow hard cheese to be sold even if contaminated with the organism.

Listeria occurs in large amounts in the sewage sludge which is spread on farmland. It also thrives in silage, especially old and mouldy silage. About forty species of animals have been found to carry it; it is even transmitted by seagulls. The bacteria are found in meat, but the cooking usually kills them. In the future we will have to find speedier methods of identifying the organism and improved systems of hygiene in food production plants in order to control the organism as far as possible. *Listeria* is a killer. Unless stringent measures are introduced, it may become a greater threat in the food of the future.

Until 1978, nobody realized that *Campylobacter* was one of the commonest causes of diarrhoea. It was the work of the Belgian microbiologist J.P. Butzler that uncovered its extent. The reason is simple: bacteria are usually cultured in conventional incubators, grown in petri dishes. *Campylobacter* will not grow in air, and never appeared on the culture plates. It needs a special atmosphere, with only a trace of oxygen if it is to grow. Butzler was the first scientist to recognize this crucial fact.

Since 1981, *Campylobacter* has been responsible for more cases of diarrhoea than *Salmonella*. The number of cases peaks during the summer, perhaps because of people handling infected meat for the barbecue, and also at Christmas, when poultry is in almost every Western kitchen. Although it is a widespread organism, *Campylobacter* infections usually require no special treatment. The diarrhoea and pain usually last for two or three days, but the patient is well again within a week. The infection usually clears up without long-term problems, but is painful and debilitating while it lasts. Some patients even develop arthritis. Patients need to drink plenty of water to counteract the dehydration due to fluid loss. In the most serious cases, erythromycin and cipromycin are suitable antibiotics.

Campylobacter is everywhere. It has been isolated from wild birds, and more than half the samples taken routinely from chicken-farms are positive. In many poultry farms all the birds are infected, and there may be as many as 100 million bacteria per gram in the birds' gut contents. Of chicken liver samples, three-quarters are typically found to be contaminated with *Campylobacter*. The bacteria are so common in nature that we cannot hope to eliminate them. However, they cause a widespread disease with unpleasant effects, and research is needed to find ways of controlling them. If consumers realize that they are likely to be contaminated by foodstuffs, they will not buy them. In the future, a single much-publicized outbreak from a named source could prove disastrous to the industry.

The bacterium *Eschericia coli* is named for the German bacteriologist Theodor von Escherich (1857–1911) who first discovered the organism. *E. coli* is the chief bacterial inhabitant of our bodies; one-third of faeces are solid masses of *E. coli* cells. Usually the organism is helpful, but some strains have been incriminated in widespread cases of food poisoning and even of meningitis.

Bacteria can conjugate sexually and exchange genetic material through a connecting tube that forms between two cells. In 1982 a cell of *E. coli* seems to have picked up a couple of toxin-producing genes from another bacterium, type 1 *Shigella dysenteriae*. This is one of the most dangerous bacteria that causes dysentery. The organism is recognized through the effect of its toxins on cultures of monkey cells, which break down and die. The tissue cultures are of vero cells, originating from African green monkey kidney, so the toxins are known as verotoxins.

The presence of these genes transforms the relatively innocuous *E. coli* into a potential killer. Although most recover completely from the infection, some suffer kidney damage which can be permanent, and can kill. The best-known type of verotoxin-producing *E. coli* is O157:H7 and this is the organism for which we usually test. There is another type, O26:H11, for which people don't usually search and which can consequently be missed. It is clearly possible that we are not casting our net sufficiently wide. The food scientist of tomorrow needs to abandon the specific search for *E. coli* O157:H7 and look instead for all toxin-secreting *E. coli* cells.

We know these as 'verotoxin-producing *E. coli*', which is abbreviated to 'VTEC'. For the food of tomorrow, we need to ensure that VTEC bacteria are properly controlled. The term VTEC is more relevant than *E. coli* O157:H7 and is a lot less long-winded. In the early 1990s, only a small percentage of cattle in the Western world were contaminated with VTEC; now most of them are. This makes it important for those who handle meat – at home as well as in larger kitchens – to ensure that fresh meat is never allowed to contaminate other produce. The meat shelf in the fridge should be wiped down with dilute bleach solution, and so should the hands. Since the heat of thorough cooking kills the organisms, VTEC on the surface of a piece of beef will be killed in the roasting oven. Problems arise particularly in hamburgers, for the mincing process carries bacteria from the surface deep inside the product. If the middle of the meat remains rare, then the organisms can survive and, indeed, multiply.

Hamburgers were behind a VTEC outbreak in the United States in 1993 which resulted in 732 victims across four Western states (including California). Of the 195 hospital cases, 55 developed kidney failure and four children died. Three years later, an outbreak involving unpasteurized apple juice produced in California resulted in over sixty illnesses and one death. Meanwhile, in a Malawi refugee camp, 20,000 people were stricken with a severe bloody diarrhoea with VTEC as the culprit, and in the summer of 1996 10,000 people were infected in a Japanese outbreak. Thirty of those victims became critically ill, of which half died. There have been several outbreaks in Scotland, one of which resulted in ten deaths and the butcher being taken to court. In the modern world, VTEC is everywhere. One case even resulted from a four-year-old boy petting a goat in a zoo.

There is no specific treatment for these infections. Severely ill patients are given support in hospitals, their symptoms being relieved as far as possible, and most eventually recover. For the future, we need to concentrate on VTEC (rather than merely on type O157:H7). In 1990 it would have been possible to isolate cattle carriers and eliminate the infection, but it is now so widespread that this desirable aim might be impossible. However, the publicity that follows an outbreak and the extent of human suffering means that all meat outlets, particularly hamburger cafés, are under continued threat from legal action.

Conquering Future Infections

Food already presents us with significant new risks of infection and in the decades ahead, other organisms will emerge. Bacteria like staphylococci were wiped out by penicillin, when it first appeared on the market during the Second World War, but antibiotic-resistant organisms pose a real threat to the future of agriculture and to the food-production industry. Whole industries could be wiped out.

Our first line of defence must be prevention. When we do encounter organisms, whether they are well-known bacteria (like *Salmonella*) or newly-emerging types (like VTEC), we will have to develop tests for them as a matter of priority and then find ways to eliminate them from the food chain. There has been one classical case of the 'sweep it under the carpet and ignore it' mentality; BSE in British cattle. The causative agent of this strange disease is believed to be a contagious malformed protein called a prion that contains no genetic material. After the familiar categories of germs, like bacteria and viruses, this is a new form of disease agent which we still find it hard to understand. I have discussed this saga in detail elsewhere (*BSE: The Facts*, London, 1996), and we will review it later, for it clearly carries lessons for the future.

If we are going to encourage a new precautionary attitude, schools need to reinstate hygiene training. A generation ago, we merely had to be careful not to introduce pathogenic organisms into our food. Washing the hands with soap after visiting the lavatory was the hallmark. Now the position is very different: we should regard all meat as potentially contaminated with dangerous bacteria. Scrupulous care will be required in the commercial and domestic kitchens of the future. Some changes have already been introduced into schools. The trading of lunchtime snacks between children (who often prefer what their friends have brought along) is being widely discouraged; in the US the tradition has been ended in the interests of safety.

One innovation I propose is a separate handbasin for a disinfecting wash. In the kitchen of the future we should find an extra washbasin solely for decontamination. I can see the disinfectant manufactures rubbing their hands with glee, confident of selling still stronger-smelling concoctions. In reality, the best disinfectant is also the cheapest – bleach. Hypochlorite bleaches destroy all bacteria. A concentrated solution of bleach (though not, of course, recommended for domestic use) is one of the few known means of inactivating the causative agents of BSE and CJD, which resist heat treatment and all normal disinfectants. In the kitchen of the future, I foresee a bowl of weak bleach solution always on standby, ready to decontaminate infected hands and implements, and a new form of disposable towel which is dedicated to people disinfecting their hands.

Do not imagine that vegetarians are free from these health hazards; many of the newly emerging pathogens can be transmitted by ready-prepared salads. Some (like *Listeria*) can even grow near freezing-point, which prevents us from relying on the refrigerator as a means of controlling their increase. Food must be seen as a potential source of infection. It is the only safe way to proceed in the future.

Allergy or Intolerance?

Chapter 2 has touched on food intolerance in the case of dairy produce. The problem is increasing. We tend to speak of such conditions as food 'allergies', a term coined in 1906 by a Viennese paediatrician, Baron Clemens von Pirquet. However, not all forms of intolerance are allergic in nature. Lactose-intolerance is an example of a non-allergic response, but that doesn't stop patients saying they are 'allergic to milk'. (Some people abuse their diet by an obsessive intake of one type of food, and recently one milk addict died at the age of 26; his fat-clogged arteries were due to his drinking five gallons a week for twelve years.)

Up to 10 per cent of children suffer from some form of food allergy, but the tendency decreases with age, so that only about 2 per cent of adults are allergic to food. Some of the most common allergies in children, which most of them grow out of, are eggs, wheat, soy and cows' milk. Allergies to peanuts in both children and adults are now well known in the West, but in Japan (where a good deal of soy sauce is consumed) there are higher levels of allergy to soy. Adults are also often allergic to seafood.

An allergic reaction can often be nothing more than red eyes, a runny nose, or a touch of wheeziness. The patient's cells have reacted to a trace of a protein in the food, or present in the immediate environment. Since there is no protein in pure oil, it is claimed that those who are allergic to peanuts may be unaffected by the oil from the same source, though I think that they'd be brave (or foolhardy) to risk it.

The most severe allergic reaction, anaphylactic shock, can kill through suffocation. One person in the USA, for instance, dies from anaphylaxis every few weeks. A follow-up survey of thirteen children suffering severe allergic reactions to food showed that six died, and seven had been close to death. In each case, the parents had been aware of their children's allergy, but the food which caused the attack had been presumed to be safe. Asthma has been found to be a common factor in children with food allergies, and it is now recommended that children should be watched for at least four hours after exposure, because they can still become severely ill.

For a true allergic reaction we recognize several key stages:

- The allergen, even if in the smallest amounts, must be present in or on the body. Unless it is present in the cells, no reaction can take place. The immune system responds only to molecules of a specific size range, and only one kind of molecule normally comes into this range: proteins.
- Other molecules, the haptens, also generate an immune response. They are smaller molecules than proteins, so they have to become attached to a protein molecule before they are activated.
- The protein which is the allergen has to be detected by the B lymphocytes (B-cells) in the body which produce antibodies. The antibodies are protein molecules that can fit on to an allergen and inactivate it.
- Each B-cell makes its own specific antibody, depending on the type of intruder to which it is responding. There are millions of different B-cells in the body that are ready to produce anti-bodies.
- The five main categories of antibodies (immunoglobulins) are: IgG, IgA, IgM, IgD and IgE. Each group is used for a different purpose, to inactivate bacteria or block the action of viruses, for example.
- In the case of allergies, the body produces the antibody immunoglobulin E (IgE). This was discovered in the 1960s by two Japanese immunologists, K. and T. Ishizaka, a husband-and-wife team.
- The normal result of the immune response is that the anti-bodies bind directly to the appropriate unwanted complex to neutralize it.
- IgE does not work like this. It attaches one of the projecting side-branches of each antibody molecule to one of the body's numerous mast cells, found in the blood. Another side-branch then attaches itself to the allergen.

- When this occurs, the mast cells start to break down. Each contains histamine, which floods through the body producing symptoms including swelling, redness and cramps.

An extreme allergic reaction causes rapid inflammation. Adrenalin is the most reliable first line of treatment, and people with severe allergies often carry a syringe of adrenalin ready for emergencies.

People who are highly allergic can respond to the smallest imaginable stimulus – one little boy died after someone opened a bag of peanuts in the same room. Just a few molecules were enough to trigger a severe response. In 1999 a 24-year-old accounts manager from Sunderland University in the north of England died from anaphylaxis after sampling a little food at an Indian restaurant. Although the menu noted which meals contained nuts, she tried a little from her friend's plate by dipping in a piece of fried potato. She died in his arms in the ambulance on the way to hospital. The previous year, a British businesswoman had died after eating dessert in a restaurant. Unknown to her, it contained tiny traces of nuts. Tragically, she had lent her adrenalin kit the day before to a friend who had just been confirmed as having nut allergy, and there was nothing that could be done to save her life.

It is now recognized that about one in every two hundred children in the West is allergic to peanuts, up to half the 10 per cent of children who suffer from food allergies. This is a disturbingly high total. Peanuts (actually a kind of underground bean) are a good source of protein and of healthy oils, and the peanut industry is a large international commercial concern. In extolling the virtues of their product, the pro-peanut lobby plays down the risk by emphasizing that only a small percentage of people suffer from the allergy. But this might mean 40 or 50 youngsters in a school of 2000 – two classes full.

There have been campaigns to make schools peanut-free zones. The same logic would apply to airliners, where peanuts are routinely handed out and where medical emergencies are best avoided. All planes now carry resuscitation kits containing a supply of adrenalin, but it is claimed that a total peanut ban is the only way to travel safely.

In my view, this would be a mistake. The banning of peanuts from specified environments gives a false sense of security. People can feel that someone else has taken charge, which removes individual responsibility. It also does nothing to lessen exposure from the everyday world, and that's where serious problems will continue to occur.

The risk to health, and even to life, will doubtless result in some notable legal cases. In the long term, I foresee the development of a vaccine. Scientists at the Johns Hopkins University have already narrowed down the number of possible peanut allergens from a group of 30 proteins to a short-list of 7, which would facilitate this. The alternative would be to modify peanuts genetically, so that the potentially lethal antigen is no longer produced by the plant. If this happens, then we may see peanuts rendered safe for everyone to eat.

Questions of Intolerance

Food intolerance is a different matter. There is not the risk of death through anaphylaxis, though the range of complaints now linked by a reaction to specific foodstuffs (or groups of foods) is becoming extensive. They include some of the symptoms of allergy:

- Disturbances of the stomach and intestinal tract, such as stomach ache.
- Irritable bowel syndrome, Crohn's disease and ulcerative colitis may result from intolerance.
- Chronic rhinitis (inflammation of the nasal passage) or sinusitis; in some cases the nose lining can be eroded.
- Skin complaints, including itching, eczema, hives and adult acne.
- Joint and muscle inflammation, including pain in the limbs and rheumatoid arthritis.
- Headache or migraine.
- Chronic fatigue, mimicking 'yuppie 'flu' or chronic fatigue syndrome.
- Asthma.
- Anxiety or depression.
- Pre-menstrual syndrome.
- Hypoglycaemia (weakness due to a drop in blood sugar).
- Sleeping disorders.

Sometimes a substance produces a marked effect in a minority of people, whilst producing a pleasurable (milder) form of the same response in the majority. The methyl xanthene compound we call caffeine is an example. In small amounts it can act as a stimulant, in larger doses causing hyperactivity, headaches and insomnia. Many people react adversely to gluten, present in wheat flour. A gluten-free diet could be tried by anyone with chronic bowel problems.

A large group of compounds, grouped together as vasoactive amines, are known to have a marked effect on some people, whilst being innocuous to the rest. Susceptible people experience flushing and feel unwell when they eat foods containing these compounds. The following table lists the main examples.

Vasoactive Amines	Source
Histamine	Fermented foods
Octopamine	Citrus fruit
Phenylethylamine	Chocolate
Serotonin	Bananas
Tyramine	Cheese, pickled fish

People with symptoms of intolerance as previously listed, might like to try eliminating foods systematically from these categories to see if there is an improvement. In the future, people will be tested for such intolerances and will bear a record of these on a smart-card. The use of swipe-terminals technology will make it easy for people to select foods that agree with them. Physical complaints due to food intolerances, within a few decades, will largely be eliminated in the Western world.

Chapter 5

The Genetic Engineer: Saint or Sinner?

If you want to see something profoundly unnatural, consider farm animals and crops. Sheep and pigs, horses and cattle, like wheat, barley, oats and rye are all new, modified organisms that are the result of human intervention. Modern maize has needed humans to help it reproduce for thousands of years since it was first bred by Native Americans from wild *teosinte* grass. Traditional farmers have been producing new animals and plants by cross-breeding for ten thousand years. The development of wheat and the farmyard pig were tremendous advances carried out by our prehistoric ancestors, and we should commemorate their pioneering achievements with pride and a certain astonishment. They put modern research completely in the shade.

Genetic engineering offers a more radical way of manipulating characteristics, as it will allow traits that would never have occurred in hundreds of years of selective breeding to be introduced in a single generation. It also allows us to take a characteristic (like frost resistance) and transfer it from one species to another. Yet the achievements of the geneticists truly have been modest by comparison with our ancestors: a cauliflower, a bulbous-eyed goldfish or a Pekinese dog are gross distortions of nature, and nobody in the modern world has ever come close to creating anything so far-fetched.

Sometimes genetically modified crops are labelled 'Frankenstein food' by the press. That's not an appropriate description. Frankenstein created his monster by grafting limbs and other pieces together, and this technique has long been used by gardeners to great effect. Most fruit trees, and large numbers of ornamentals in parks and gardens, are produced by grafting one strain on to another. Usually you can see the line of the graft, just like the seam on the monster's neck. Trees with several varieties on the same rootstock are a popular novelty item in many private gardens, and these much-loved plants truly are Frankenstein creations.

Genetic modification is a very different matter. The discovery of methods to cut and splice DNA means that we now have the capacity to identify individual genes, and we can harness natural phenomena in the living world to import modified genes into new hosts. The genes could be attached to viruses, for example, which would carry them into the host cell as they enter it. Some plants, the monocotyledons, are not amenable to the conventional biological methods of transferring genes, so the modified DNA is actually fired into the host cells with a device like a specialized shotgun. Tiny metal particles are coated with the modified DNA and shot into the cells at high velocity. Using these methods, we can insert genes into the chromosomes in reverse sequence (so that they no longer work) or we can take genes from one species and introduce them into another. The result is an organism whose genes have been modified to provide the desired characteristics.

Genetic modification is inevitable. Like electric power, road transport, or computers, it is a facet of the future and the public will gain little by campaigning to ban this potentially rewarding technology. Properly applied, it could offer us so much. This does not mean that we should meekly accept it. A *laissez-faire* official attitude to genetically modified organisms, or GMOs for short, could threaten the environment, and might even have implications for human health. For example, the pollen of modified crops had already been shown to be poisonous to monarch butterfly larvae. Although you could argue that this was an artificial experiment, the suspicion remains that there might be similar effects on earthworms. It is the earthworm population that helps maintain the condition of soil, yet no research seems to have been done on this topic.

Rather than unquestioningly accept genetic modification, we need to examine it and ensure that its power is used for good, and not for ill. People need to be empowered so that everyone can participate in the decisions that underpin the introduction of these crucial new techniques.

It is important that the people are involved, because the industry itself is expanding at an extraordinary rate. The major American companies are making vast investments. Novartis, formed by the merger of Ciba-Geigy and Sandoz in 1996, have budgeted $600 million in agricultural genetics over the decade. The European Association for Bioindustries reckons that the sector will be worth $285 billion by 2005. This is a huge new industry, and it will have pronounced effects on us all. In future, we will need to control it.

Haste and Danger

The problem with genetic manipulation is not inherent in the discipline, but lies in the secrecy and the small-mindedness of the businessmen who have been trying to make money from it. Genetic manipulation has been introduced hastily and with every sign of shortsightedness.

Although there had been several key stages in the development of new GM plants since the 1970s, the first major commercial step was the introduction of genetically modified soya. The new varieties were modified in order to make them tolerant of the herbicide glyphosphate, which could therefore be used to control weeds. The modified plants were unaffected by the spray, which meant that the farmer could easily obtain a monoculture of soya bean plants and little else. This is supposed to reduce the amount of herbicide used in spraying fields, but in practice the converse happens. Because the crop is not going to be affected, no matter how much spray is applied, farmers often tend to overdose their crops. As the American *Food Magazine* reported in the autumn of 1999, there can be higher levels of pesticide residues in genetically modified soya than in conventional supplies.

What happened to the new GM product was regrettable and highly problematic. The producers ensured that the soya product from the GM plants was blended with that from conventional crops. The proportion of soya from the genetically modified plants was initially small, but led to all soya being regarded as suspect. The producers recognized that nobody would know which they were buying, and the public had to face that – since they were already buying the product – discrimination would be impossible. In 1998 it was already estimated that one-third of maize and soya products from the United States were from genetically modified crops. Once the North American markets had accepted these products, the producers were convinced that they had achieved a *fait accompli*. The reaction of Europe to GMOs was far more immediate than that of the American people. In Britain there was an increasing sense of outrage, with protests that led to the three-year moratorium on the commercial planting of genetically modified crops.

In my view, the reason for the sense of suspicion was triggered by the tragic saga of BSE, or mad cow disease, which had sensitized the population to possible problems inflicted by ill-considered changes in the production of food. The entire population of Europe was affected by the ban placed on the export of beef from Britain, following the outbreak of BSE, and the threat to human health that was revealed.

It has to be said that I have never understood why the European authorities, claiming to believe that British beef was too hazardous for mainland Europeans to consume, still permitted it to be eaten by the British. European officials may say that they did not regard the risk to human health as worthy of a total ban in Britain; but, if that is the case, there was equally no reason to ban its export to other countries.

What is true is that two major sectors have been blamed for the tragedy – the farmers and the scientists. Neither was responsible. The farmers were never given any indication that the production of their animal feed had been changed, and that there might be new risks of infection. The truth was kept from them throughout, with repeated parliamentary comments that beef was wholesome and there was no risk to the public. One can hardly blame the farming community for believing that this was the truth.

Meanwhile, the scientific community were concerned from the outset. Rumours that no scientist expressed doubts about the potential problems are falsehoods perpetrated by officials with a vested interest.

The paid members of the official advisory committees may have been issuing bland statements whenever the media asked questions, but academic biologists were concerned from the start. All the published reports, even those commissioned by the British government, were couched in terms that revealed the sense of concern felt by scientists.

On 4 June 1987 a note published in the *British Medical Journal* by two doctors first postulated that mad cow disease might be transferable to the human population. That same year Sir Richard Southwood, the Linacre Professor of Zoology at Oxford University, was appointed to chair a committee to inquire into BSE, which had been officially recognized the year before. If we read some of the extracts from his report, the sense of concern is clearly conveyed:

- Paragraph 5.3.1: With the very long incubation period of spongiform encephalopathies in humans, it may be a decade or more before complete reassurance can be given.
- Paragraph 5.3.5: Because the possibility that BSE could be transmitted orally cannot be entirely ruled out, known affected cattle should not enter the human food chain . . . .
- Paragraph 5.3.6: It is a reasonable assumption that were BSE to be transmitted to humans, the clinical disorder would closely resemble CJD. Depending on the route of transmission, the incubation period could be as little as a year . . . or several decades [and] specialists . . . should be made aware of the emergence of BSE so that they can report any atypical cases or changing patterns in the incidence of disease.
- Paragraph 9.2 (Conclusions): From the present evidence, it is most unlikely that BSE will have any implications for human health. Nevertheless, if our assessments of these likelihoods are incorrect, the implications would be extremely serious.
- Paragraph 10.4 (Summary): Concerned at the remote chance that this new infection could be transmitted orally to man, we recommend the destruction of carcasses of cattle with suspected BSE and prohibition of the use of milk from such cows for humans. These recommendations have already been acted upon.

The Southwood Report, February 1989

The bulk of this report was written during 1988, just one year after BSE was officially recognized. The Committee's judicious warnings were set out with perfect clarity. They emphasized that:

- They could give no reassurance for ten years or so.
- They could not rule out the chance that BSE could be passed on in food.
- The disease could take over a decade to appear.
- Meanwhile, specialists should watch out for it.

In the event, it was not until March 1996 that the British authorities admitted that the disease had spread to humans. Young British people had tragically begun dying (though in very small numbers) of a disease that was distinct from Creutzfeldt-Jakob disease, and was in my view much more like the sinister *kuru* of the cannibal tribes of Papua New Guinea. The cause was a mixture of ignorance, commercial greed on the part of the feed manufacturers and an unwillingness on the part of the government to do anything that might interfere with big business. Faced with this lethal combination, I think it was very natural that the European response to GMOs was going to be ultra-cautious.

Late in 1999, a committee of scientists in Brussels agreed that British beef no longer posed a threat to human health. In spite of an edict from the European union, France promptly refused to import supplies, and so did several of the German *länder*. The British scientists were insisting that the beef was safe, and newspaper campaigns on both sides took up the cause. Was there any reason for the recalcitrance? It remained the fact that the retail sale of beef on the bone was still banned in Britain, on safety grounds, which did little to mollify objections. Furthermore, there were still about 2000 cases of BSE in British cattle during 1999.

Before we condemn the French for being totally unreasonable, we can try to put the problem another way. Just supposing that French Golden Delicious apples had been associated with a disease that destroys the human brain, and 2000 of their trees still contained the infection: how many Golden Delicious apples do we imagine Britain, Germany or the USA would be importing from France today?

Learning Lessons from BSE

Although European sensibilities were influenced by our knowledge of BSE, there are few direct comparisons between GMOs and the BSE debacle. We know far more about the way genes function than we knew of the prions that are believed to cause BSE.

The introduction of genetic engineering is following on a century of detailed research into how genes work and what they do. Chromosomes had been discovered by Karl Nägeli in 1842, and were given their present-day name as long ago as 1888. Hugo de Vries, a Dutch botanist, first recognized mutation in 1886 in plants of the evening primrose, and at the end of the century a paper written by the Moravian monk Gregor Mendel showed how inheritance worked through cross-breeding peas. His paper was first read in English at a meeting of the Royal Horticultural Society in London in May 1900, and a special meeting of the Institute of Biology on May 8, 2000 was to mark the centenary of that crucial revelation.

Mendel pointed the way to modern genetics. First, he recognized that discrete genetic characteristics were passed on from parents to offspring and showed how dominant and recessive genes could operate. Secondly, he cheated. Mendel was confident he was right anyway. Having recognized the likely mechanisms, Mendel wrote up extensive tables of results which – we now know – are too good to be true. He invented the figures he needed. There are resonances in modern research, where a sense of conviction can prevent people from following proper procedures.

Sometimes we have carried out procedures that seemed fine at the time, but which acquire an aura of unimaginable irresponsibility with the benefit of hindsight. The story of what became known as 'monkey glands' provides an example. In the 1920s in France, doctors treated aging wealthy men by injecting them with testicular tissue from chimpanzees. It seemed such a simple procedure: the injections provided the source of testosterone which the elderly patients required. In today's world, the whole idea seems fraught with potential disaster. The transplantation of tissues in this way can pass viruses from an animal to the human host. It has even been argued that the 'monkey-gland' treatment of the 1920s may have been the mechanism through which HIV was introduced into human society.

If future food scientists must bear any single thought in mind, it is that the quick fix does not always succeed in the long term.

A Warning for the Future

There should always have been serious doubts about the efficacy of introducing GM crops into the market without public consultation.

- First, and most important, it liberates genetically abnormal organisms into the environment with the minimum of supervision. New drugs are put through a detailed appraisal before they are released for human consumption, whereas health foods are subject to minimal controls. The new category of genetically modified organisms has been subject to far less testing than one should reasonably expect.
- Secondly, it denies choice to the consumer. In a nation like the United States this should have been considered a fundamental interference with individual rights. A lack of labelling contradicts the principles of freedom of information and freedom to choose.
- Thirdly, the main decisions are taken by commercial interests which depend upon shareholders whose prime concern is to make money.

In Britain, questions have been raised over the impartiality of the regulatory authority. Government ministers rely on the conclusions from the Advisory Committee on Releases to the Environment prior to the field trials. Critics have pointed out that the majority of the members of the Committee have links to the food industry, and stand to benefit from the new business that such trials might help to create. No proposed field trial has been rejected by the Committee.

In 1999, the British government announced a tightening-up of the regulatory authority. The response demonstrates how timely were the concerns, for a list of proposed environmental experiments was hastily rushed through to beat the deadline. Meanwhile, calls from the industry to make field trials secret have been consistently rejected by the European authorities. The sites for planting genetically modified maize and oilseed rape have to be announced in advance.

Large corporations are rushing ahead with ill-considered techniques. They have used antibiotic-resistance genes as markers, heedless of the fact that resistant bacteria present us with a threat of colossal proportions. They have released pollen from genetically modified plants into the environment, claiming that it will not spread. People who suffer from hay-fever when a considerable distance downwind of an oilseed rape field know that pollen can travel a long way. Some wild species of mustard and its relatives are reported to have picked up the modified genes. In a climate of opinion that is cautious about interference with the mechanisms of life, it is clearly counterproductive to press ahead with research that does not embrace public consultation. If ever we need an object-lesson in how not to introduce new technologies, the story of GMOs is the ideal example. We cannot blame the press; it is the industry that has been at fault.

Trials and Disappointments

In our attempts to harness biology for the future of agriculture, our disappointments are more numerous than our successes. It has proved impossible to imitate photosynthesis, for instance. Green plants manage to take the simple compounds of water and carbon dioxide, utilizing solar energy to combine the atoms to form carbohydrates. Nobody has ever fully imitated this process, on which all life depends. A blade of grass or a cup of pond-slime can perform a miracle which is still inimitable by science.

For the genetic engineers, the grail of nitrogen-fixation remains unattainable. Some plants (notably the legumes, like peas and beans) acquire communities of bacteria with a rare ability to take nitrogen from the air and process it into nitrate. They produce fertilizer for the host plant out of the air. The transfer of the nitrogen-fixing genes to crop plants like wheat could allow us to farm very poor soils. It seems so basic – but still we cannot match it.

Plants as a Source of Pesticide

The notion of a plant that produces its own pesticide is either exciting (if you are a genetic engineer) or deeply worrying (if you are a sceptic).

Natural plants are a rich source of pesticides. This is how they resist pests. Bitter cassava has long been a highly successful crop in the tropics because it lies in the ground for prolonged periods, ready to be disinterred when the time comes to eat it. The reason it survives intact is because it is full of cyanide. Unless it is ground and boiled prior to being prepared it will kill the consumer.

Plants are a source of alkaloids and carcinogens. The cabbage family is well known to contain cancer-causing chemicals, while many of the leguminous plants are poisonous too. Less well known is that runner beans are dangerous until cooked, and there is a disease in the countries of the Eastern Mediterranean known as neurolathyrism which is caused by eating chick peas. Many plants, such as laburnum, lily-of-the valley and cuckoo-pint, whose seeds and berries are highly attractive to children in their play, are extremely poisonous.

The reason spices have strong tastes is because they contain a vast range of poisons. They are poisonous in order to avoid predation, but the strongly tasting compounds they produce have become prized as ingredients for a human diet. When safety analysts test spices for bacteria, they have to dilute them before they try to start a culture. If you test pure spices, no bacteria will grow: they are inhibited by the spicy-tasting poisons that the plant produced as it grew.

I see little reason to prefer strains of lettuce, for instance, that are inherently poisonous to pests. A pesticide we spray on to a plant is one whose properties are reasonably well known, and the use of which has been approved by a regulatory authority; plants that are naturally poisonous are riddled with strange toxins about which we still know little. Plants that produce their own internal pesticides can be a hazard to those who eat them. During the 1980s, a strain of celery was grown in the US that contained raised levels of psoralens, which were believed to be useful in controlling pests. In practice, the psoralens were found to be changed by the action of sunlight and became irritants to human skin. Farm workers harvesting the celery soon developed a serious skin rash. A natural variety of potato, Lenape, was withdrawn in the US in the 1960s when it was found to secrete dangerously raised amounts of solanidine glucosides, which are poisonous.

Many natural foods are potentially harmful. There are at least 500 chemicals in coffee, for example; and, of the twenty or so that have been investigated, all are potentially carcinogenic. Excuse me while I pour a cup, just to get over the revelation.

Organic farming frequently encounters such philosophical problems. It is very laudable to aim at the production of vegetables and fruit without the use of liberal amounts of chemical sprays and fertilizers, but if the result is a crop that is replete with its own internal chemical armoury against the world of insect pests then the safety of the produce remains questionable, no matter how organic it may be. There will be a need in the future to balance the risks from organic crops with the benefits that genetic engineering could yet provide.

So far as current commercial crops are concerned, the major GMOs are these:

- Pesticide-producing maize from Monsanto.
- Herbicide-resistant maize produced by AgrEvo.
- Maize from Novartis, resistant to both pests and herbicide.

There are questions over these. One issue is the matter of herbicide resistance. It is being claimed that the actual amount of herbicide used on farms growing GMO crops is much the same as that used on conventional farms. There would thus be no saving on cost, nor any reduction in environmental damage. Meanwhile there is a disadvantage in that the crop (because it resists the herbicide) is the only plant in a field that can survive, so we end up with a monoculture of the crop and nothing else.

This is a foolish way to farm, and I do not say so out of some idealized position of protest. Part of the problem is that the weeds in fields sustain wildlife, and maintain the rich diversity of nature. Without healthy soil, crops are growing in a hydroponic system. All the soil does is hold the plants upright, while the crop is fed with artificial fertilizers. In the real world the huge communities of microbes in the ground process nutrients and offer them to the roots. All this is threatened when plants are grown in a monoculture. We might as well stick the seeds into gravel and feed them fertilizer from a factory.

This approach to farming damages wildlife and locks the hapless farmer into a multinational monopoly to supply their livelihood. Each farmer has to purchase seed from the multinational, has then to purchase the patented herbicide, and then go back next year for more seed. His tradition of saving seed from one year to the next no longer works, because the GM crops have a gene which prevents their producing viable seeds, although the some of the multinationals are reconsidering their policy on this. The resulting sprays that the farmer has to use kill off huge numbers of indigenous plants and threaten the microbes on which life in the soil naturally depends.

However, I predict that the consequences for the companies could be even worse. Firstly, not all the weeds are killed. Ninety-five per cent of them are destroyed, but 5 per cent survive. This means to me that, in a short while, resistant weeds will take over the fields, so the farmers will be back where they started. Not only that, but the newly implanted genes will inevitably escape. There is a view that the maintenance of a buffer zone around a test-field of GM crops prevents the pollen from escape. But you can find pollen grains in the stratosphere, and in the middle of the Atlantic. No matter how much these technologists know about genes, they have never grasped how an entire organism functions. Pollen spreads; it's like a germ (hence 'germination' and 'germ cell'). It takes more than a strip at the edge of a field to stop pollen grains from spreading round the countryside.

I predict that the resistance to herbicides will soon spread. Such herbicide-tolerant weeds are known as superweeds, but this is a misleading term. They grow no faster than normal weeds. Neither will they will spread more rapidly than conventional plants. Their resistance will, however, mean that the chemical herbicides will no longer function. We will be back at square one, as though the herbicides had never been discovered.

When that occurs, there will be nothing to sell. The commercial benefits will be gone, and nature will have taught the large commercial companies another lesson. Nothing would induce me to invest in one of these shortsighted companies. They are heading for major problems, and the future will show that the effects on them will be far more serious than anything that commerce can inflict on nature.

New Genetics, New Food

New technology is already raising ethical issues for both vegetarians and non-vegetarians. The North American winter flounder is a fish noted for its ability to resist the extreme temperatures of wintertime. The genes that confer the ability for survival may be introduced into tomatoes or strawberries to produce fruit which sustain less damage when frozen. Some vegetarians argue that such tomatoes are no longer suitable for their consumption.

Not all genetic engineering transfers genes from one species to another. One pioneering commercial example of genetic modification was when the British bioscience company Zeneca reversed a gene involved in the spoilage of tomatoes. The gene itself was unaltered, but it was inserted backwards so that it could not be read by the cell. The enzymes it produced were responsible for the breakdown of the tomato. This process is part of the fruit preparing for the germination of the seeds within, so it is an important part of the tomato life-cycle. To the farmer, however, the rapid decay of the fruit is a source of loss, whilst to the consumer it meant that tomatoes did not last as long in the kitchen. The reversal of a gene is known in the trade as antisense technology ('antisense' simply means 'backwards'). Its effect is to inactivate the gene, though there are no new genes in the end result. It is not a complete answer to the breakdown of tomatoes in store, but does mean they stay fresh and wholesome for up to twice as long.

In the future we will see a range of valuable new GMOs. We will grow coffee plants that produce no caffeine for those who need to avoid it, bananas that secrete vaccines against serious diseases like hepatitis and meningitis, and potatoes that absorb less fat when fried.

Principles for the Future

Biologists tend to view the future of genetic manipulation as promising. Quite apart from food, our knowledge of genetics offers so much.

When we consider the children doomed to die through cystic fibrosis, or the daily suffering of a physically incapacitated writer like Stephen Hawking, it is impossible to turn aside from the benefits that genetic modification will bring. The era of vaccination, the discovery of vitamins, the dawn of surgery through the development of anaesthetic agents and the provision of clean drinking water have been landmarks in the progress of civilization. Genetic engineering will be a further step in our quest for understanding and humanitarianism, and is the most promising new discipline of this era.

The public look at matters very differently when it comes to GM food. Stanley Greenberg was commissioned by Monsanto, who had been involved in a range of GM trials, to find out how attitudes in Britain were changing. He reportedly found that the percentage of people who thought that genetically modified crops were 'acceptable' fell from 33 to 25 per cent between 1997 and 1998, while the proportion who felt strongly opposed rose from 38 to over 50 per cent. Among the AB social classes the number opposed to GMOs rose from 47 to 57 per cent, and opinions of Monsanto in this same group were 2:1 against the company.

Meanwhile, a major British supermarket chain Iceland announced in 1999 that they would no longer sell GM products, and then began sourcing eggs only from producers who did not use any additives to produce golden yolks. Another supermarket chain, Sainsbury's, started advertising that all their own-brand products were free from any GM ingredients. Marks & Spencer had begun to sell GM-free produce in the spring of 1999, but by the autumn they were announcing a range of meat products from animals reared on GM-free foodstuffs. This does not augur well for tomorrow.

There is an object lesson here in how *not* to handle the public relations of science, and in the future we must address the need for public accountability and openness. We should now agree on acceptable criteria by which to establish ground-rules. In my view, there is an urgent need for a code of criteria which would enable governments around the world to assess new developments and set our controls. Among them I would propose:

- Approval for new experiments should always be sought from a regulatory authority well-versed in the subject.
- None of the members of such a committee should be in a position to benefit commercially from approval.
- No genes conferring problematical properties (e.g. antibiotic resistance) shall be used, whether as markers or otherwise, outside enclosed laboratories.
- Agents capable of transfer to wild plants, like pollen-carrying genes conferring resistance, shall not be liberated into the environment.
- The public shall be consulted about the siting and the benefits of experiments.
- All GMO products should be properly labelled and the source declared.
- Records of possible unwanted side-effects should be meticulously maintained. All such events should be investigated by an organization unconnected with the source of the product.

These are the musings of your commentator, and not a set of hard-and-fast proposals. I dare say some of these could be faulted, and yet others might be proposed by wiser minds. What I am saying is that we need an agreed framework for the future. This is the only way in which the promise perceived by science and the problems postulated by the public can ever be objectively assessed. The potential is too great to ignore – and the hazards too serious to be underestimated.

CHAPTER 6

CAN THE FUTURE FEED THE WORLD?

The farmers of the world face a challenging future, but can celebrate the fact that they already produce enough food to feed everyone. Famine is caused by a failure of distribution, rather than a simple lack of food. More problems are caused by warfare than by shortages. Ethiopia, for instance, has been a net exporter of grain, and a prominent importer of luxury goods like whisky, during periods of widespread famine in its provinces.

We have to find ways of providing an equitable supply of food that does not depend upon crude market forces. In the West it is popular to believe that the free market is the engine that drives a successful society, but the reality is very different. All too often, agribusiness is regulated from above. Prices are fixed and subsidies are used to control what farmers do and the crops they grow. There are hundreds of traditional crops that it is now unlawful to sell anywhere in Europe.

In the future, small farmers will re-emerge to provide the luxury items and the special foods that people enjoy. Food that is traditionally grown and free from the intrusion of technology will become increasingly desirable among the cognoscenti. Self-sufficient farming will increase, with direct sale through the Internet and elsewhere, and with less intervention from official bodies. Currently, subsidies that were envisaged as a way of protecting farmers in poor areas are being commercially exploited by wealthy landowners. Since farming damages the soil, we will need to return much more organic waste as compost if we are to avoid continuing desertification. Financial support in the future should be designed to support farmers when calamities occur, and not to control their lives. Instead of governments supporting big businesses, what they should do in future is to provide a buffer for the farmer in times of crisis.

World Food Production

To most of the world, a supermarket is nothing more than a distant dream. The Africans who grow French beans for the British company Tesco, for example, believe that 'Tesco' is a space-age country far away. In Third World countries you can find a cramped shanty roofed with corrugated plastic, weathered and split, with the word 'supermarket' hand-painted on a board.

The reality of the world to hungry people is very different from what we know in the West. This morning, 800 million people awoke starving. They will be equally hungry when they fall asleep tonight. This is tragedy, completely unacceptable when comfortable Western nations are replete with food, and waste huge amounts of it. Although farming seems like an international enterprise, most food is consumed in the country where it was produced. Only 16 per cent of food crosses national boundaries between production and consumption, though this amount is set to rise in the coming century.

As we have seen in Europe and the USA, a reluctance to import food from one state to another can quickly rebound in the form of trade wars. Although Americans are willing to fatten their cattle with growth hormones, the practice is disliked by Europeans. Yet British turkeys are often so bulky that as adults they cannot stand properly and they suffer from sores, facts known to very few of the public. For one nation to protest about the other seems hypocritical, to say the least.

Predictions for the future of food have often been wide of the mark. It was predicted in the West that China would be unable to expand animal feed production fast enough to keep pace with her need for meat, and that Chinese imports of coarse grains (corn, sorghum, oats, and barley) would considerably increase. International suppliers planned ahead to capitalize on the demand. It didn't happen. Throughout the 1990s, coarse grain exports from China ceased. Homegrown high-protein meal made from oilseed crops increased threefold. The production of grain went up by 15 per cent, while total agricultural production went up by over 50 per cent. In the same decade, animal products almost doubled. No nation has ever shown such a formidable increase in food production as China since 1990, through land reclamation and increased efficiency.

In nations around the world, high crop yields have either been impressive (US) or unpredictable (Russia), while natural disasters can reduce a predicted surplus to nothing (Bangladesh). Although the world is not yet approaching the biophysical limits to food production, there are global indications that growth in food production has begun to lag. For instance, food production increases did not keep pace with population growth during the 1980s and early 1990s in 49 developing countries with a population of one million or more. The annual rate of growth of global grain production also dropped from 3 per cent in the 1970s to 0.7 per cent during 1985–95, a potentially disturbing warning for the future.

We can try to put these figures into a human perspective. Of the 800 million people who are starving today, 175 million are children aged under five. About one-third of the world's infants are exposed to illness and death because they are malnourished. Those statistics are hard to visualize, so let us make it more familiar: since you began to read this chapter, forty children have died because their parents could not feed them properly. Every hour that passes, another 500 children die. The fact that one-fifth of humanity is hungry will be seen as unconscionable, and international authorities should now accept that a subsistence supply of food is the birthright of everyone.

When we plan for the future, we need to harmonize import and export regulations so that the free movement of much-needed food is unimpeded. We should set up task forces, ready to rush food to areas of devastation at a moment's notice.

The Future of Food Supplies

In was around 1970 that we began to see the 'Green Revolution' as food production started steadily to increase. By the end of the millennium we were producing more than enough food for everyone. In 1970 about 900 million people were starving – perhaps a billion in total. Little wonder warning bells were sounding. Since then, agricultural production has soared. By 1995 the amount of food produced had increased by 55 per cent. Since the world's population was increasing, the drop in the number of starving people went down less dramatically, but it did decrease to 800 million. In proportional terms, this is itself a fine achievement. It means that in 1970 one-third of the world was starving, whereas by the 1990s the number was down to one-fifth.

The problem for the future is the need to keep pace with mushrooming global population growth. In 1960 the world population was 3 billion; it passed 4 billion in 1974, 5 billion in 1987, topping 6 billion just before the turn of the millennium. It has doubled within living memory. Yet still I do not see the size of the population as an insurmountable problem. Writing of my words in a scientific journal in October 1999, the Chicago-based Professor Walter McCrone chided me gently by saying: 'The only conclusion I haven't heard him propose is that there are too many darned people in the world.'

There is a reason for this. First, reproduction rates fall as personal wealth increases. Furthermore, food productivity has been buoyant throughout the recent decades. In 1960 the index of food production per person was 83. By 1970 it reached 90, in 1990 it was standardized at 100, and by 2000 it is about 105. We can afford to be proud of this achievement.

Over the next 15–20 years global food demand will increase by 30–50 per cent, and farming will have to keep pace with that requirement. Already the signs are appearing: in 1970, Africa produced as much grain as she consumed whereas at the turn of the millennium she produced only 80 per cent of her needs. Grain production in Africa is down over 10 per cent since 1980, and 22 per cent since 1967. We will keep making reductions in the proportion of people who are starving, but it is calculated that if we continue as normal we will still have 600 million hungry people by 2010. It is a lower figure than we have today, sure, but it is still worryingly high. The challenge of the future of food is to reduce that number.

Feeding the Hungry

There are clear categories of people who go hungry. Doubtless, some are feckless individuals who could do better but simply don't trouble.

Sometimes they claim to be free and independent, while relying on a civilized community to provide them with what they require. But they are a tiny minority. Most people are not by nature scroungers, and, although some governments like to assume that everyone who is poor is too indolent to do anything about it, that view is unsustainable. In the dry deserts, and other places where food is scarce, there are young people doomed to die before they ever have the chance to use their talents. They die for the want of a little food.

The impoverished families in developing nations have low or variable incomes, few if any assets, and they lack marketable skills and advocates for their cause. Because they have no money for the system, the system has no interest in them. Typical examples are the unskilled unemployed, daily workers who own no land; livestock herders (shepherds, for instance) and small-holding farming families. They survive on the edge of subsistence, victims of drought, floods or a failed harvest. Many are in thrall to moneylenders. Some sell their children into slavery or prostitution. Their earnings are less than a dollar a day, three-quarters of it spent on food. At least 50 million of them are refugees, displaced from their homes by warfare. Most refugees are hungry.

Many suffer from disease, yet little effort goes into treating or preventing tropical diseases. The people have no money for drugs, so the international drug companies don't usually trouble to develop vaccines or treatments. A few companies are setting aside a little money for altruistic research that could benefit countries too poor to offer a profit, and we can only press for this to expand. Heaven knows, we spend enough in buying high-price, low-cost drugs in the Western world, and profits from pharmaceuticals are high.

This mindset underpins the future of food. Many people are hungry because there is no food: they have not produced enough, or none is getting through. However, and this is the crux of the matter, even if food were to become available, they could not afford to buy it. They live in areas where the infrastructure is weak, where marketing and distribution costs are high – areas which are economically marginal. There is little incentive to develop schools, hospitals, shops, factories, farms;

and many of the options that are developed are shortsighted and based on expediency rather than on long-term integrated planning.

One example is the methane-producing plants that were planned for widespread introduction to India, which seemed so organic and wholesome. Street women live by collecting cowpats from the streets and drying them on the trunks of palm trees. Each cowpat was slapped into position where it dried with its prominent hand-print, looking like a work of art. When dry, it would fall to the ground to be collected and used as fuel. Cowpat stoves were everywhere, using the droppings of the cattle that roamed the streets as a mainstream fuel supply.

The gas generator promised to change all this. Instead of burning the cowpats, a crude and inefficient process that polluted the air, they would be mixed to a broth and fermented in closed vessels to produce methane. This meant that every village would have its own source of homegrown natural gas, which was cleaner to burn and easier to control than a cowpat fire. In time, local villages had piped gas, running from the gas generator.

And that is when things began to go wrong. The cowpats now had value, for they were needed to keep the generators working. As a result, the street women started to sell them to make a little money. Since this left them without fuel, they began to cut wood from the ancient forests. As a result, areas of India were threatened with deforestation simply because gas generators had been introduced. A seemingly non-invasive and natural technology was threatening the ecosystem.

Elsewhere in the world, rainforest is being destroyed. Great swathes of the Far East have been blanketed by smoke for weeks on end as huge areas of forest are burnt. In place of the trees, grass grows, and on the grass cattle can be raised as a cash product. It offers short-term gains (by selling or burning the timber and grazing the land), but it displaces communities and denudes the land of root structures it needs to survive. Floods tear away the soil, and people drown in their thousands. Crops are lost for a season, and epidemics break out as new parasites invade the land. These areas attract little inward investment. Without investment, they remain unproductive. Yet the people are eager and bright, intelligent and industrious; they are longing for a future – while starvation condemns them to an early death. Even if there were to be economic growth, starving people are unable to take part in it. Hunger is not merely a result of poverty today. It is the cause of poverty tomorrow.

In future we need to encourage the development of short-term food-for-work schemes which allow people to make progress without the enslavement that can come from working at low rates for an unscrupulous employer. These schemes could encourage the development of infrastructure, like roads, irrigation, public amenities, schools, sewerage and water services.

We will need to introduce foods that are culturally acceptable to the varied tastes of the global populations. These should include snack foods providing vital proteins, carbohydrates, minerals, vitamins and fats. Whatever nourishing food we produce in the future, we must ensure that it is desired by the people for whom it is intended. It is not enough to distribute bread, for example, unless it is the kind of bread to which people are accustomed. Flat breads are far commoner around the world than loaves like the white, bulky bread favoured in the West: they are often thin and crisp, and of very different consistency. The nearest we get to that is the base of a pizza, which (although it originated in Italy) is really a Middle Eastern flat bread. All across Arabia such flat breads are made, though by the time you move down to Ethiopia the snacks are made with bread as thin as a crêpe.

The Naples restaurant Antica Pizzeria Port'Alba, founded in 1830, claims to have been the first to make a pizza in a wood-burning oven. The origins of the habit are now lost in antiquity, so nobody can challenge the claim, but the flat bread known by the remarkably similar name of pitta remains common throughout the Balkans and dates back for centuries. Indeed, I postulate that it was the Phoenician traders who came in search of tin and silver who brought the idea to Cornwall, where it survives as the Cornish pasty, and, although made with pastry rather than dough, is similar to the daily food throughout the Arab world. In the future, whole meals will be presented in a readily marketable form, like the pasties or pizzas that began as a portable meal for working people.

Women Face the Future

The burden of chronic hunger is passed on from one generation to the next like a hereditary disease. In a typical peasant family, it is the women who are fed last, the belief being that the (hardworking) father needs the best food. This means that the one person who needs the best diet is actually getting the worst. A poorly nourished woman becomes small and thin. During pregnancy her unborn child strips her of nourishment for its own metabolic needs, so she becomes still weaker. The child is born seriously underweight, and underweight babies cannot catch up with large ones. A good diet during childhood will enable a small baby to catch up part of the way, but most of the disadvantage remains throughout life. The smaller adults give birth to tiny babies again, and so the weakness is inherited.

If any sector commands attention for the immediate future of food, it is the women. Seventy per cent of the hungry people around the world are female, yet it is the women in those communities who produce 70 per cent of the food. Although the total number of hungry people has gone down since 1970, the number of starving females has doubled in that time.

Meanwhile, anaemia afflicts women around the world – menstruation causes a regular loss of iron. For many years ferrous sulphate was used as a supplement. Some years ago I took part in experiments in which we drank iron salts labelled with radioisotopes in order to see how much of the iron was retained by the body. It turns out that the most popular iron supplements were not retained by the body at all. These days, ferrous fumarate is believed to be more readily absorbed by the body.

We have so much to learn about female nutrition. For example, many women in the US take calcium supplements, drink milk, eat cheese, etc., yet still suffer from osteoporosis after the menopause, even breaking bones. The exact cause remains a mystery, and osteoporosis has become a scourge of Western female society, with painful and time-consuming consequences. In West Africa, on the other hand, nobody ever drinks much milk; cheese is unknown, calcium supplements unheard of and disease is rife. Many of the women have eight children or so, which depletes the mother's calcium supplies. Yet osteoporosis does not occur.

We still do not know why this might be, although much research has been devoted to the condition. It has been suggested that an alkaline diet encourages the deposition of calcium whereas acids (like the phosphoric acid in cola drinks, for example) leads to their dissolution. This is unlikely, for the body's internal levels of acidity are separately regulated and are not primarily influenced by the diet. The hydrochloric acid in your stomach is strong enough to burn a hole in the carpet, for example, and its strength does not depend upon acid in your diet.

Fortunately, the introduction of hormone replacement therapy, HRT, is reducing the occurrence of osteoporosis in Western women as they get older. In planning foods for the future we need to look carefully at a range of such research. There are so many nutritional problems we have yet to understand, and they hold the key to the full empowerment of women.

Land Use in a New Millennium

Farming is one of the main means of land management around the world. The landscapes of Europe and America, like the settled areas of Australia, are the result of farming. The land area of our planet measures sixteen times the size of the United States – 93 million square miles (240 million square kilometres). Arable land makes up one-tenth of the total, meadows and pastures a quarter, forest and woodland about a third. The remaining one-third includes frozen wastes, deserts, mines and towns. In Asia, 80 per cent of the available arable land is already under cultivation; there is very little unfarmed land left in the world. The rest is too steep, wet, cold or dry to be productive.

We can free up meadows for crop production and we are already seeing rainforests stripped out for new farmland. People in the West, where we have long since stripped many of our own forests, deplore this trend. Forest clearance and habitat destruction are currently running at about 618,000 acres (250,000 hectares) every week, which is why such large areas of the tropical East spend long periods each year under a dangerous smoky fog caused by the burning of ancient woodlands a thousand miles away. In the USA you will often hear statistics on the rate at which tropical jungles are vanishing, but rather less about the rate at which natural American wetlands (a comparable resource) are being daily destroyed.

By reclaiming meadows, there is more land that we can utilize for food production. The supply, however, is finite. The future will need to find how to utilize land effectively, rather than to hope for an unending supply. How far can we extrapolate into the future? Estimates from the United Nations and the World Bank are that the population will reach 10 billion in 2060. Rates of economic growth, currently around 3 per cent, are expected to fall to one per cent per year by the middle of the century, and cereal yields are expected to increase by an annual average ranging from 0.9 per cent (developing countries) to 0.6 per cent (developed nations) over the same period. The most important change must be the abandonment of tariffs, and the freeing of international food exchange. Food is too important to be regarded as a crude means of making money for distant shareholders.

Future Climate Change

Current climatic calculations are based on the world as it is. How will we stand in the future, with a climate modified by human intervention? Our burning of petrochemicals is already having accelerating effects on the global climate, and little serious effort is being made to improve matters.

We can model how crop yields might vary with temperature. As levels of atmospheric carbon dioxide and methane increase, the greenhouse effect will trap increasing amounts of heat. At the same time, there will be raised levels of carbon dioxide in the atmosphere, which should in theory cause plants to increase their rate of growth. CO_2 is a very rare gas in the atmosphere (less than four parts per thousand). It is surprising that plants manage to obtain sufficient supplies as it is, for CO_2 is their sole source of carbon and plants are largely built of carbon compounds.

However, the theory of unalloyed benefit to plants does not translate into practice. For decades people have concluded that an increase in CO_2 will promise a bonanza – a natural fertilizer in the air. After the Earth Summit in Rio de Janeiro in 1992 there were many reports that the raising of CO_2 would aid the growth of plants, but it is not so simple. For some species, the increase is a stimulant. Oranges will produce more flowers and an increase in fruit yield, but only if irrigation can keep up with demand. However, staple crops (including soya and maize) do not respond to raised CO_2 by giving higher yields. Although wheat yields would increase slightly, average grain size decreases.

An increase in atmospheric CO_2 has the effect of reducing plant res-piration. Currently the bulk of food production is concentrated into relatively few crops. Wheat, rice and maize alone compromise about 85 per cent of world cereal exports, while soya accounts for two-thirds of trade in protein cake. This will pose problems if climate change produces marked effects in a particular crop. Meanwhile, attempts to control levels of pollution have been disappointing. European nations are keen to aim at a 15 per cent reduction for CO_2 emissions by 2010, though other nations (including the United States and Australia) have been slow to agree. Australia has recently overtaken the American rate of increase in atmospheric CO_2 pollution.

Changes in temperature could be beneficial. If average tempera-tures increase by 2°C, there would be corresponding increases in productivity. In maize and rice the rates would increase by about 8 per cent, whilst productivity in wheat and soya could increase by between 10 and 15 per cent. However, they could also have devastating effects.

First, the implications are not the same all around the world. Assuming rainfall stays the same, an increase of 2°C in Canada would lead to an increase in wheat yield, whereas exactly the same change in Pakistan would lead to a fall in yield of about 12 per cent. When the temperature rises, the growing period is shorter because the plants are growing faster. This gives less time for fruit to form. If the tempera-ture is raised, evaporation rates could be expected to increase, and this could put plants in greater danger of water shortage.

Even more disturbing is the danger of changes in the circulation of the seas on which much of the civilized world now depends. The unpredictability of El Niño in 1997–98 is reckoned to have cost at least $37 billion, but worse could be in store. Much of Europe depends upon the Gulf Stream to moderate the climate. The northward flow of tropical seawater is greater than a hundred River Amazons, and it is drawn along by the sinking of vast water masses as they cool in the Arctic oceans. If the climate were to become too warm, this cooling and sinking could stop, and the Gulf Stream would cease to exist. If that happened, cities like Liverpool in England would have a climate like that of Spitzbergen or the Hudson Bay. The heating of the tropics would mean that the lush train forests of Brazil could become dry savannah within 50 or 100 years.

Global warming is largely due to the loading of the atmosphere with carbon dioxide. Industry and the private motor car are forcing the levels steadily upwards, and already we can see that the global climate is more energetic than it was. Can we expect to see a paradigm shift, where people want to drive less? Practical experience tends to make one feel the answer is 'no', for the United States resolutely refuses to adopt any of the widely accepted protocols that limit car use, and remains the greatest per capita polluter in the world.

Yet there may be hope for the future. When the warnings about the effect of CFCs (chlorofluorocarbons) on the ozone layer were first mooted, nobody imagined it would be easy to banish those aerosols. In fact, Western industry quickly rallied round. Once the risks were accepted, controls were instituted and these damaging compounds were eliminated, even though their effects will linger on for many decades yet. If that sort of responsible change could be brought to bear on other ingrained habits that threaten the atmosphere, then we may yet hope for an effective answer to climate change.

Certainly we need to fear a refusal to change our ways. One-third of the world's human population lives on land that is liable to be inundated if the seas rise. Most of the great cities are near sea level. Many nations would disappear if the sea were to rise by a mere half a metre, yet in the past we know that the sea levels have risen and fallen by over

100 metres. If we are going to maintain the modern world, then concerted action for the future is urgently needed.

Surviving the Future

We need to plan for the future thoroughly, holistically and globally, so that we can produce enough food until the world population stabilizes later in the century. Population increase is a function of poverty, since wealthy people see less need to spend money on large families and prefer to enjoy themselves. At least in the short term, wealth creation results in increased selfishness.

In poor countries, children are the only way a family can increase their possessions. They come free. They can work for the family as they mature, and care for the parents when they are old. For these reasons, rates of reproduction are inversely proportional to personal wealth. When women are seen as childbearing housemaids, reproduction rates remain high; when they are seen as equal partners and as emancipated, freethinking individuals, childbirth rates go down. In the Indian province of Kerala, women still occupy a backward role in society, and birth rates remain high. In states like Uttar Pradesh, by contrast, the place of women in society is more equal and their dignity is higher, so birth rates are far lower. As modern civilization spreads, population increases become more manageable.

It is estimated that world cereal production, which totalled 1.7 billion tonnes in 1990, will top 3.25 billion tonnes by 2060. Even that would still leave 500 million people hungry. Unless we act promptly, the number could be even higher. I believe that a coordinated campaign could eliminate hunger altogether within twenty years, and I cannot see any reason why we should not set out to do so. President John F. Kennedy set a man on the moon by the end of the decade as his target for the 1960s. It was a formidable ambition, but the commitment of minds brought it about.

We now need to recognize that we must now focus on ending the brutality of global starvation. The military might of the nations, coupled with our technology and full-hearted support, should be used to target the end of world hunger within a decade. I am certain that it could be done, and there could be no finer aim for the immediate future. The use of our military might in peacetime draws attention to the need to help provide food for hungry people after a major disaster. We should establish an international Food Force that can be dispatched urgently to disaster zones. We have seen heavy earth-moving equipment rushed to earthquake areas for some years, but little sign of help for the millions left destitute and hungry after extreme weather.

The cyclone that struck the Indian state of Orissa late in 1999 left millions in a state of starvation. The winds blew at 180 mph (290 kph), driving 20-foot (6-metre) waves across the land. Two million homes were destroyed, and 20 million people had their lives disrupted. Yet, a week after the two-day storm had ended, the fate of 1.5 million villagers was still unknown. Modest supplies of rice were being dropped by army helicopters, but people were becoming ill with gastroenteritis caused by drinking contaminated water. An international organization should have sent immediate supplies of water purification tablets, dried food and electrolyte concentrates for the sick and starving. In the future we will surely no longer tolerate the death through neglect of vast populations of our fellows, whilst we sit at home and watch the affair unfold on television. We are often told of the need for armies to train to the peak of fitness and efficiency, and here is the perfect role for them in the future.

It can no longer be acceptable that our fellow human beings are trapped in wretched conditions, profoundly hungry and ill through neglect, while billions are paid to equip military machines that do little. The main task of armies is feeding themselves; now it is time for them to offer the experience to others when disaster strikes. We are witnessing the increase of energy in global weather systems, so further major disasters are going to become more likely, and not less.

We need to establish Food Force as a matter of top priority. This is the first time in history when we truly could solve these pressing problems, and the future is at hand.

CHAPTER 7

NEW IDEAS FOR A
NEW MILLENNIUM

Supermarkets, like other large enterprises, exist to make as much money as they can. Designers of supermarkets aim to increase spontaneous purchases. The right paint is chosen for the walls and the lights are carefully selected. The whereabouts of specific items is changed from time to time, to the irritation of the regular shopper, but because it is believed that this increases overall sales performance. Research is even showing the extent to which mood determines what people buy. The British chain store Marks & Spencer find that indulgences are most often bought on Friday evening, with comfort food preferred on Mondays and specialities midweek. On Mondays, therefore, they will feature ready-made meals, such as cottage pie, or sausages and mashed potatoes; at the weekend beef Stroganoff or Spanish-style chicken are prominently displayed. This belief is supported by research at King's College, London, which emphasizes that carbohydrates have a calming effect, whereas proteins increase alertness and agitation, and that people who exercise a lot tend to consume more fat and sugar.

Personal preferences are studied in detail by supermarkets, which seem to be designed to overcome one of the conspicuous results of contemporary education – an inability or unwillingness to communicate. In this modern environment, there is no need to communicate with anybody. Because you act as the shop assistant (unpaid, at that) the supermarket arranges that the flow of money from your pocket into their vaults is not impeded by the need to articulate anything to anybody at any time. It is the ultimate example of dumbing down.

The public who buy these products from a supermarket often imagine that they are the customers. This is incorrect. It is the *shareholders*, not you, who are the customers. Supermarkets give as much of your money as possible to their shareholders and senior management, while their task is to drive down the price of everything they buy. They go to suppliers and offer them a steady outlet for their produce. To the farmer it seems like a godsend. The insecurity of the marketplace is avoided at a stroke – all they have to do is supply a single outlet and the future is secure. It may be secure, but it isn't very profitable for long. Once the supplier is locked into the system, the supermarket drives down the price they pay until the suppliers are at their wits' end.

This is not to keep the shop prices low. Quite the contrary. Prices go soaring up the moment there's the hint of an increase in wholesale costs, but they are much less likely to come down when a supplier's price is cut. The British beef farmers found that they received half what they were paid before the BSE crisis, though the price of meat to the public has remained pretty static.

In any event, the prices charged by supermarkets are not as low as you think. Everywhere you go there are signs reassuring you of the generosity of the supermarkets. They allow you to keep change in your pockets; they have your low prices uppermost in mind; they do whatever they can to help make life more convenient. That's what they claim. They are much less likely to tell you how they are screwing down the supplier's costs, in order to make a greater profit out of you.

A Resurgence of Shops

In the future, individual shops, friendly, warm and welcoming, will begin to reappear. What's more, there will be no interminable queue at the till, and your money will go to someone in the community who is very likely to spend it locally.

For many items the shop comes to the house, another pointer to the future. As I was writing these words, we had a visit from the mobile grocery store. It calls here every Friday. Just to be fair, I noted the prices and then called to three nearby supermarkets to see how they compared. Two stores had signs up offering super low prices on bananas. The mobile grocer's prices were lower. The peppers were cheaper, and the eggs from the mobile store were larger than those in the supermarkets, but much the same price.

The other retailers who call at our village home include a butcher marketing his own meat, a fishmonger direct from the fishing harbour, a general grocery, even a specialist dealer in tea, coffee and confectionery. This idea of home delivery may seem quaint and old-fashioned, but it is very likely that this is how matters will develop in the future. With Internet shopping available to so many people in the Western world (and spreading exponentially to the rest), direct access to suppliers has never been easier. Although supermarkets are flashy and space-age, a future generation will eventually tire of traipsing round the endless aisles. In the future, we will order supplies on-line, and rely far more on door-to-door delivery and sale.

It is important to recognize how people are attracted to the novel built environment, like a child going to a theme park. People use cars to drive the shortest distances, claiming that they really don't like to walk if they can avoid it, that they are in a fearful hurry, and that the car brings its own liberation. They will drive a few yards to buy a news-paper or a packet of cigarettes. However, put them in a novel built environment and they will walk for miles. Not only is the modern supermarket visually attractive, but so, for example, are airports. The busy executive will walk here, though she would use her car for far shorter journeys in more familiar surroundings.

If you want to control the vogue for greed and exploitation, then start using local suppliers. There's no greater pleasure than handing over money to a local supplier who helps make life easier. Now that global capitalism is in disarray, it would make sense to support local businesses. The supermarkets may be the black holes in the firma-ment of commerce – but it's the local suppliers who are the stars. With the Internet to hand, nobody is too isolated to remain out of touch if they are selling a product you wish to buy.

For all the future emphasis on nutritious foodstuffs, this is going to be the era of deregulation and we will see a return to local production and sale. The public may not know what they want, but they do want what they know. Traditional handmade breads, long-lost home-cooked dishes, carefully prepared meals that take time unavailable to the modern working family will all give the small operator a key role in the supply of future foods. Many indigenous traditions can be encouraged. We will see people farming unlikely crops, like elder whose flowers make an appetizing cordial, and whose berries make wine. In the rainforests the harvesting of indigenous crops, like Brazil nuts, will be encouraged. Rare grains, unusual fruit (including obscure varieties, some of which cannot legally be sold in today's restricted market) will come to the fore, and specialist providers will flourish in the future.

The Eclipse of Meat

Meat will disappear from most Western diets. The pressure will come from concern over meat-borne infections ranging from VTEC to BSE, coupled with the growing realization that growing crops is a much more efficient use of land for producing protein than raising livestock.

There will also be an increase in humanitarian concerns about animal welfare. Such farm animals as are still reared for consumption will face a very different life from the inhumane conditions of the twentieth century. The increasingly urban population will find the idea of eating animals distasteful, reinforced by a feature of modern life in the West – the city farm. These little farms set up on waste ground to familiarize children with 'farm' life in reality make them sentimental about baby animals. Children raised in the countryside are far more matter-of-fact about eating chickens, lambs and cows.

Composite Foodstuffs

Much of the food of the future will be produced like pet-food pellets, created and combined to provide a perfect balance of vitamins and minerals. Currently we have breakfast cereals with added minerals and fortified with vitamins; tomorrow's special foods will contain additives covering everything a healthy body can require. As children's choices for lunchtime snacks turn away from junk food, a highly concentrated source of vital dietary constituents will take its place. The final presentation of these nutritious foodstuffs will be in a form acceptable to the consumer: as chocolate snacks, for example, or as fillings for sandwiches.

Future foods will feature synthesized meat substitutes, like the fungus protein of today. Cultured fungi and other microorganisms will be textured and fortified with vitamins before being artificially flavoured prior to packaging. Artificial flavouring will become widespread. It is a curious paradox that many people turn to vegetarian foods because they wish to turn their backs on what they see as the chemically adulterated foodstuffs of the modern world. As a result, vegetarians are among those who eat most synthetic and artificially flavoured foods. Mycoprotein made from fungus cultures can be textured to make a very palatable form of non-meat food. Yet, not only is the fungus grown and harvested through entirely unnatural means, but high technology is used to texture the product, and synthetic flavourings are used to complete the process. The acceptability of such healthy foods in the future is assured. We already produce beef-flavoured crisps, for example, which have never been near beef, and which are even marked 'suitable for vegetarians'. In future we will be able to enjoy a range of savoury vegetarian foods that would please the palate of the meat-eating customer.

Currently, the widespread application of the by-products of the meat industry (like gelatin) means that there are meat derivatives in unlikely places – chocolate mousse and liquorice allsorts, for example. We are thus in a strange situation where many non-meat-containing products would be acceptable as meat substitutes, and could be eaten by vegetarians; whereas there are many ostensibly vegetarian products that do, in fact, contain products of the meat industry.

We need to watch for opportunists who will try to find ways around future regulations just as fast as safety measures are introduced. The management concept that lies behind current attitudes to safety tries to balance cost with expediency, and has given rise to the acronym BATNEEC – Best Available Technology Not Entailing Excessive Cost. The problem is that many companies regard human casualties as a part of the equation, and they cost compensation into their operating strategy. For many organizations, the preferred approach is based on the principle of CATNAP. Watch out for it. The term means Cheapest Available Technology Narrowly Avoiding Prosecution.

The Meal in a Pill

There is nothing new in the idea of a complete 'meal in a pill'. It was a Victorian concept, and indeed featured in a story written by the great Canadian humorist Stephen Leacock prior to the First World War. He wrote of a greedy child who, at a Christmas meal in the future, took everyone's condensed meal tablet and ate them all. Leacock described how the swiftly reconstituting meal in the child's stomach led to his rapid expansion until he exploded, the walls being littered with fragments of the young glutton .

Future food technologists will be well placed to create concentrated foodstuffs which rectify the known deficiencies in the diet. We already have compressed tablets that provide a daily supply of vitamins and minerals; There will be a product in the future that will include all the known constituents for health. Such foods will need to be supplemented by a diet of traditional foodstuffs, for there must be important trace components in food about which we still know nothing. We should remember that only in recent years have some brands of formula milk begun to incorporate the long-chain fatty acids which are now believed to be important in the development of brain cells.

This is the problem with synthesized foods, where the components are assembled according to a fashionable recipe. A rare element may be a crucial component of some enzyme system, for instance, and if it is not included in the formula then it will not be available for the consumer. The consumption of everyday conventional foods, however, allows such rarities to remain part of the diet. The people of unsophisticated societies enjoy a vast range of dietary inputs. Thus, the hunting of an animal by prehistoric people would mean that many of its organs were consumed, together with much of the gut contents. The range of raw materials for the body would be extensive. Modern foods, for all their technical sophistication, cannot match this broad spectrum.

Tastes of the Future

Flavourings will be the hallmark of future food production. Potato and rice snacks are already in production with synthetic flavours (chicken or prawn, cheese and chives, etc). These are popular choices for vegetarians and slimmers, and chemists have analysed the tastes of typical foods so that they can be synthesized in a pure form. Thus, benzaldehyde confers the taste of pineapple, whilst a derivative of this molecule, 4-hydroxy 3-methoxy benzaldehyde, has the taste of vanilla.

These synthetic compounds are purer than those produced in nature, but this essential purity produces problems of its own. Natural foodstuffs contain a range of flavours that create a subtly different sensation. Recently my wife and I were served ice-cream at an old-time cinema and immediately turned to each other to comment on the depth of the taste. I kept the carton, and found afterwards that there was no synthetic flavouring in the list of ingredients, but only the extract of the vanilla orchid seed-pod. The subtlety of the taste derived from other components in the extract, and these can be missing from products flavoured with synthetic additives.

Food technologists are now proposing to engineer 'designer vegetables', like meat-flavoured tomatoes or prawn-flavoured potatoes. Banana-flavoured Brussels sprouts are even under consideration. These may sound abhorrent, but it has always been the setting of unexpected flavours in juxtaposition that has led to the development of new foods. Dill and pickles, for example, are a strange combination, as is horseradish and roast beef. Sweet and sour dishes have become internationally popular in the space of a generation or two. The blending of such disparate tastes would have been anathema to our grandparents, although a few generations earlier Europeans had often blended meat with fruit. Mincemeat, for example, was originally made with minced meat and dried fruit. Future foods will create unexpected blends of tastes set in contrast, and novel products, like fruit-flavoured vegetables, will probably be welcomed by many consumers over the next decade or two.

Fungi, Familiar and Unfamiliar

We eat more fungi that you might at first imagine. Blue cheeses owe their appeal to growths of fungi and the fruiting-bodies of some kinds of fungus are a traditional item of the diet. The mushroom is perhaps most familiar in Britain, but the edible boletus is popular in Europe, and Oriental shii-take mushrooms are increasingly popular in the modern Western kitchen. We have for centuries harnessed yeast to produce a huge range of foodstuffs from bread to beer and wine. Much of the surplus yeast from the brewing industry is salted and sold as a paste in England and Australia (though curiously not in the United States). These brown savoury spreads are sold under registered trade names such as Marmite, Yeastrel and Vegemite, and there are *aficionados* of each brand. They are rich in proteins and vitamins of the B group, but they are an acquired taste. Advertisers often say, 'You either love them, or you hate them'.

In the East, the culture of microbes for human consumption is a time-honoured tradition. In Japan certain bacteria grow in colonies under the soil surface on the slopes of Mount Asama, an extinct volcano, and are harvested and made into a soup named *tengu*. The filamentous alga *Nostoc* is grown in ponds and is known in China as *to-fa-tsai* ('hair plant'). The alga *Spirulina*, which could be grown in vast amounts using simple vats as culture vessels, is already becoming a popular new food supplement in the English-speaking world, and the recent development of mycoprotein in the West, made from fungi grown in vats, is a rational extension of the idea. The product, once texturized, has a similar mouth-feel to meat. Culturing microorganisms offers a highly efficient means of producing high-protein food supplements for a hungry world of the future. A few hundred square kilometres in the Arctic (which would provide the cold water to cool the cultures) would theoretically be sufficient to feed the world with all the protein we need.

Science and the Future

Chemistry allows us to generate molecular structures that can have specified effects on the body. Although we challenge these because of their synthetic nature, it is important to bear in mind what I have said on the unnatural nature of our traditional diet. Many of those foodstuffs would stand no chance of approval were they invented today. Butter, sugar and cheese would never have been permitted were they present-day developments. Many novel foods, like mycoprotein, are arguably safer even though they are the creation of modern technology. As discussed in Chapter 2, there is nothing 'natural' about bread and butter, yoghourt and cheese and, just as ancient technology gave us those foods, future technology will offer us others.

The problem facing us (and it is a real problem) is that traditional technologies took thousands of years to perfect, allowing many undesirable consequences to be eliminated, whereas modern technologies are often launched without time for adjustment. For centuries, innovations were made by enthusiasts. In the modern world they are being exploited by businessmen with little interest in the long term. For the enthusiast, it is satisfaction that matters; for modern managers the drive is for money. Few of today's innovations are properly thought through. A full consideration of long-term consequences must guide our decisions in future. Most people in authority seem to regard short-term as tomorrow, medium-term as next month, and long-term as next year. That has to change. Short-term means next year. Medium-term means next decade. Long-term means next century at least. Even that may be too restricted a view, for that's when our grandchildren will be around. It is their world that our current policies will affect.

We will need to rethink the role of science in public affairs. There is a vogue for asking scientists to pronounce on whether or not something is 'safe'. This has never been the remit of science, and the modern version of management that tries to seek assurances in this way is giving us bowdlerized and unrealistic expectations. Scientists are searchers for the truth, or at least they were always supposed to be. In the modern world they fall foul of competing pressures. Scientists are sometimes so insecure in their positions that they tend to mouth conclusions that will please the employer, or at least not embarrass him. They are also expected to make moral decisions rather than present evidence. The Royal Society of London, one of the world's greatest academies, always used to publish a statement that it could not offer opinions on any issue, for that was not the purpose of science. Years ago that disappeared, and today official bodies are always being asked to issue statements on safety issues. This has never been the function of science. In future we should confine scientists to divining the reality of complex situations, and not demand that they pass subjective judgments on matters of personal choice.

We often hear of 'the scientist and the public' as two camps. What we need to grasp is that the scientist *is* a member of the public. No scientist is better than anyone else at reaching a conclusion over safety; what they can do is determine the facts of the matter and present those as fairly as they can. Scientists regularly berate the media for 'getting things wrong', but at least the newspaper reporters can be forgiven for doing it: they don't know all the science, and they have to sell copies. Scientists are also liable to confuse the issues, and – with all that training and prestige – they have far less excuse for doing so.

In the future, we need to put science back in its place. Science has never had opinions, and had better start remembering the fact.

CHAPTER 8

NEW FOODS

The future of food production will be very different from the past. Our present-day habits are encouraging diseases to spread. Slurry from animals with chronic diarrhoea is hosed down with impunity, and often ends up in watercourses that feed the human population. If you watch a herdsman at work as he inoculates his animals you aren't likely to see him changing the hypodermic needle, using a fresh one for each animal. This will all have to change if the spread of disease is to be avoided. Food is now a leading source of infection and in future we will have to regard all meat, in particular, as a potential source of disease.

Our conventional attitudes to farm animals are inhumane and exploitative. Traditional breeding techniques are responsible for much animal suffering. Broiler chickens and turkeys are already under great stress because they grow so quickly that their legs can barely support their own body weight. Eighty per cent of broiler chickens now suffer leg deformities. Large turkeys are often unhealthy and deformed by the time they are slaughtered. Pigs under stress can develop pale, soft exudative meat (PSE) where the product is soft and unmarketable.

Worst of all, many farm animals throughout Europe are routinely dosed with antibiotics, which can act as growth promoters. Europeans object stridently to the use in the USA of growth-hormone pellets inserted into the neck of beef stock, to make them grow faster. One can find plenty of objections to the use of growth hormones, which can cause animals to suffer from the excessive muscle they produce. There have been concerns expressed over the effect of the hormones on the consumer, though scientific evidence does not seem to substantiate that. But the liberal use of antibiotics in Europe is certainly far worse. Growth hormones may affect the animal that receives the dose, and they might even turn out to have effects on individual consumers, but indiscriminate antibiotic use can affect the whole of society.

When penicillin was discovered in London in 1928, it heralded the antibiotic era and by the 1960s we had a growing range of such drugs available to treat a range of infections. From the start, resistant organisms have appeared. Now there are serious threats from antibiotic-resistant organisms. They include some recently identified strains of *Staphylococcus aureus*, which lives on the skin and cause a range of conditions from boils to septicaemia. In recent years people have died in Western hospitals from forms of TB that are resistant to antibiotics.

As far as we can tell, the use of antibiotics in agriculture has not directly caused these particular problems. The common factor is in our attitude: we have overused antibiotics with little thought for the consequences. Dosing farm animals with antibiotics because they accelerate their growth rates cannot be acceptable practice. The more antibiotics there are in the environment, the more resistant strains we will find. More to the point, as resistant strains emerge, the greater becomes our need for new antibiotics to cure sick people. It is a race we have almost lost. Vancomycin is now the antibiotic of last resort, and some strains of *S. aureus* have now been discovered which are resistant even to vancomycin.

Controls on growth hormones in rearing cattle are all very well, but in an era when the overuse of antibiotics in animal feed is widely sanctioned, the attitude seems hypocritical. In future we will need to control all such substances. The health of the whole human population hangs upon these decisions, and prompt action will be required.

Healthy Foods for a Healthy Future

Foods that confer health have begun to attract a range of names in recent years. They have been branded as probiotics, functional foods or nutraceuticals. This last term is redolent with the blending of nutrition with the world of pharmaceuticals. The marketplace for these products is growing. In the inaugural issue of a new journal to mark the acceptance of the term, *Nutraceutical World*, launched in December 1998, the Direct Selling Association published a breakdown of the main sectors. Their categorization of a market sector worth $22.21 billion by 1997 was as follows:

- Leisure/education   4.7%
- Home/family care   31.9%
- Personal care   27.0%
- Services, etc.,   18.7 %
- Wellness products   17.7 %

In round figures, this suggests that one-third of direct sale was for homecare products, a quarter for personal care products, and one-fifth for 'health foods' and related products. The fact that the leisure and education sector amounted to no more that one-twentieth of the total reminds us how important is personal health to the public. Retail sales of vitamins, supplements and minerals are expected to reach $12,300 million in 2001, a compound annual growth rate of over 13 per cent.

What Will be Missing from the Future Diet?

Until the present era, special diets were restricted to several clearly defined categories. The medical categories were for such conditions as high blood pressure, diabetes, allergy and food intolerance. Personal dieting was confined to bodybuilding or weight loss. Now we have special diets which restrict proteins or fats, keep categories of foods separate, or allocate 'points' for each meal. The aim may be to reduce weight, it may be to reduce specific degenerative changes (like wear in the joints or cellulite formation) or to cut the chances of a heart attack.

Dieting for weight loss is due to be replaced by dieting for health. In the recent past, there have been many fad diets that have little scientific evidence to support them. Some have restricted foods, or placed them in groups; some claims have even been made that consuming certain foods will consume more energy than the food itself contains. The only way to lose weight is to consume less energy than we utilize. Fad diets are rarely successful.

In the future we should move away from the tendency to control people's diet by frightening them. We are always being warned that specific foods are going to kill us, rather than enjoined to relax and choose food because it does us good. The incidence of anorexia nervosa in young women in the First World is disturbingly high, yet it seems to fit perfectly well with what they are taught by the middle-aged generation. The media emphasize that thinness is an ideal (film stars and models with even a hint of fulsomeness are teased in magazines as if they were chubby). Health columnists warn that being over-weight predisposes us all to heart attacks and coronaries, strokes and cancer. One consequence of this is that the self-image of young women is often at variance with reality. A survey in 1999 of British children by the Schools Health Education Unit showed that 60 per cent of teenage girls believed they needed to slim, whereas only 10 per cent were actually overweight.

The diets of the future should encourage people to enjoy their food. The mainstream food categories of the present-day world will largely disappear. Millions of people are unwell because of reactions to dairy produce and to the gluten in bakery goods. At present these foods are regarded as essentially healthy and natural, but in future we will recognize that neither category is natural, and for many they are a hazard.

Public concern over levels of disease germs and saturated fats in meat will cause a move away from beef and towards pale meats, like pork and poultry. However, the high incidence of *Salmonella* and *E. coli* make even this half-way-house seem to me like a temporary phase. Rearing poultry in open fields will offer some respite to the public conscience, but since chickens developed from woodland birds, it is probably more natural for them to be in a large orchard or even a shed. It may be that in future the humane method of farming poultry will be something closer to present-day battery farming, but without the cruel overcrowding.

It is said that there are financial disincentives to abandoning present-day conventions (intensive farming, routine use of antibi-otics, etc.) but the future will show how shortsighted is this view. The public reaction against inhumane farming will lead to a major down-turn in demand for intensively produced meat products. Farmers need to realize that the losses sustained by modernizing their industry

and making it more humane will be far less than the economic consequences of a major market collapse.

Some moves towards a healthier awareness of the implications of a
traditional diet have already brought about drastic and unforeseen
changes in marketing. A generation ago, skimmed milk was a waste
product of the dairy industry, left after the cream had been removed.
Now, skimmed milk is the largest sector, yet it is a product that was
poured down the drain in living memory. Since skimmed milk has to
go to the factory, before it is returned to the store for sale, it is never as
fresh as whole milk. Semi-skimmed milk is simply a mixture of fresh
milk and skimmed, and can be the least fresh of all. Whole milk
remains one of the freshest items of food available in the shops.

The production of skimmed milk brings in its wake a new problem:
namely, what to do with all the cream. The answer makes a mockery of
the switch to skimmed milk, for the cream which is removed in the
processing is simply added to other items of the daily diet. Whenever
you look at the contents of prepared foodstuffs you are likely to find
cream. It is present in dressings and spreads, soups and sauces.

As a result, most people who buy skimmed milk are still consuming as much cream each week as their parents. Look at the ingredients
of ice-cream – the chances are it contains skimmed milk and cream.
Factories devote huge amounts of energy to separating the cream from
milk and pack them in different containers for collection by different
companies; yet in a nearby town the food manufacturer is ordering
separate supplies and mixing them together again. This is a wasteful
and dishonest use of resources which future policies must discourage.

Change for the Worse

Genetic modification, properly introduced, could yet bring us many
benefits in the future. We will face a new green revolution as crops
with higher productivity are introduced on to land that we cannot currently cultivate. Many healthy foodstuffs for the future will rely on
simpler technologies. We will see a new range of intermediate new
foods which will meet customer demand for a health-giving diet. Eggs
that are high in polyunsaturates and the health-giving linoleic acids

(see Chapter 3) can be produced by chickens who have not been genetically modified, but who are themselves fed a specially selected diet.

There will be an expansion of 'ethnic' foods, which will lead to a great increase in fusion cuisine where culinary cultures are combined. Much of the interest in cultural fusion centres on the acquisition by Western cultures of Oriental traditions. In the last twenty years there has been a rapid increase in the number of Eastern and Greek restaurants in Australia, for example, to parallel the Vietnamese, Korean and Japanese establishments in the United States.

The greatest change has been seen in Britain. Once known for its unadventurous cookery and a drab, traditional cuisine, Britain has become a gastronomic mecca. The finest British dishes have always been enjoyed by international gastronomes. Roast beef, for example, has become part of the European vocabulary as 'rosbif', and the great British breakfast is a feature of executive hotel restaurants in many nations, though, as I found out, when hosting our Channel 4 television series *Food for Thought*, that only about one per cent of British households apparently make a traditional cooked breakfast.

In Britain, Indian restaurants have appeared in every town. It is an idiosyncratic term, for almost all of them are run by Pakistanis. An evening out at the pub is very likely to end in an Indian meal. If not, it will be a Chinese takeaway, which has brought a new range of tastes to the palate of a conservative nation.

However, the pattern of modernization and change works both ways. In Italy, France and Germany the traditional methods of cookery are rapidly being lost. Traditional cookery is time-consuming: vegetables being chopped by hand to unimaginable fineness, meat being sliced, marinated, specially arranged . . . there is no time for such lengthy preparations in a busy world. The reason French cuisine is rightly celebrated around the world is because France maintained a feudal society with a vibrant peasant sector for longer than the other nations of Western Europe, and it is among the peasant farming communities that such traditions survive. We think of Italian food as being based on pasta and pizza, but French cooking was derived from the splendidly varied cuisine that the provinces of Italy can boast.

Modern recipes and preparation techniques fly in the face of tradition. Chop suey is unknown in China, as tikka marsala is foreign to India. Many modern meals are gross distortions of the real thing, and in future we may find consumers wanting to experience truly authentic dishes. *Spaghetti bolognese* is a dish of international renown, yet few commercial cooks know how it should be made. The essential ingredients are finely diced carrot and freshly-sautéed chicken liver. Not only that, but it is tagliatelle (not spaghetti) which is the favoured form of pasta in Bologna. In the English-speaking world, *chilli con carne* is a popular meal, and everyone expects to see red kidney beans as the most recognizable ingredient. A Mexican family making this dish would never use red kidney beans, however; the traditional vegetable base has long been chopped potato. In future we may see the re-emergence of real food, cooked for nostalgia in a time-honoured fashion.

In Australia there are Aborigines who can still recall how their forebears used to make the earliest kind of bread in the world. Grass seeds were separated by hand, ground on sandstone rocks, mixed to a grey paste and cooked in the charcoal ashes of a fire. A few very old people can remember exactly how it was done. In a few years they will all have died, and a tradition dating back to the roots of our culture will be lost forever.

I saw another ancient tradition vanish in India. In the old quarter of Delhi stood a shop selling a rare and traditional curry sauce. At the end of each day a bowl was kept as the cauldron was cleaned, and it was added to the mix for the next day. The process dated from about 1200 AD, so the modern customer was eating a tiny amount of food that was eight hundred years old. The old quarter was demolished one night to make way for a multi-storey car-park more in keeping with a forward-looking nation, so (at 3 o'clock in the morning) the ancient shop was bulldozed to the ground. Many traditions that are regarded as effete and old-fashioned by their practitioners become prized items for the gourmet, if only they have the chance to survive.

As the traditions disappear, more modern food habits are spreading. In the developing nations of the Pacific Rim, which have bequeathed new tastes to the West, in China, even in post-communist

Russia, there are new hamburger restaurants imported from the United States. The hamburger, which, as the name suggests, arose with the traditional ground-meat dishes of Hamburg in northern Germany, is a present-day symbol of Westernization. Although the saturated fat content, plus the risk of VTEC infection, do not encourage me to view hamburgers as a long-term proposition, at the present time all developing nations want to have hamburger joints. Similarly, fish-and-chips, a traditional British takeaway meal, is increasingly popular across the globe.

Tastes from Around the World

In the future, we will discover exotic foods that currently play no part in the conventional international diet. Some of them will have to change in order to keep pace with modern practices. Coca-Cola bears little relationship to the original product, for example. When first made it was flavoured with the cola palm, indeed the characteristic shape of the bottle is based on a bud of the palm-tree from which the flavouring was derived. The stimulant in the present-day product is caffeine; originally it was cocaine.

Where will we find future inspirations? Apart from the United States, the rest of the Americas have much to offer. The Arawaks who were original inhabitants of the Caribbean gave us allspice and cassava, one of the most widespread staple foods around the world and, as we have seen, a potent source of cyanide. Further Caribbean influences could come from *callaloo*, a créole soup made of leaves.

Meals made from vegetables in all their variety will increase in popularity. And what an array of vegetables await discovery! Some of our most popular foods come from the deadly nightshade family, the Solanaceae, also a potent source of poisons, like many vegetables. Tomatoes were cultivated in Europe as an ornamental garden plant for centuries before anyone realized they could be safely eaten (they were originally known as 'love apples'). Cookery with that other familiar member of the family, the potato, was pioneered by the Mixtec and Inca people. It was the Incas who invented freeze-dried potatoes, *chuño*; they still have over a hundred potato varieties unknown in the

developed world, many of which will be suitable for commercial cultivation. For thousands of years the Incas have made *llapingachos,* potato and cheese cakes, and *arepas,* corn buns.

From the Incas we have the bright yellow colourant annatto (from the plant *Bixa orellana*). The European tradition for a bright yellow colourant is saffron, obtained from the anthers of crocus flowers (not those of the autumn crocus plant, which is a poisonous plant). Both saffron and annatto are widely used for their coloration, but both were originally also enjoyed for their subtle taste.

Less subtle is *cassareep,* a seasoned bitter cassava juice used by the Incas to season meat dishes. In Brazil there is widespread use of *vatapá,* a shrimp and almond sauce for seasoning chicken. The Maya in Central America perfected the production of *masa harina,* a form of cooked maize flour, and *guacamole,* from the Aztecs, is already internationally accepted. This small area of the world is rich in new gustatory experiences for us in the future.

Moving westwards, we find the Pacific islanders who perfected the barbecue, mastering the technique of cooking a whole pig in steam. From Samoa comes *palusami,* made from taro leaves and shrimp, while Fiji can offer a delicious vegetable named *duruka,* like pale asparagus. Indonesian *sambals* have offered further inspiration to supplement the traditions of the West: pounded chillies, garlic, onions, tomatoes. The Netherlands' occupation of Indonesia popularized *nasi goreng* in the West: dishes made from fried rice, vegetables and fish, etc., now known as a 'rice table', or *rijstaffel.*

Japanese daily cookery has traditionally featured frugality, and their tofu and creamy bean curd are already popular items in Western supermarkets and delicatessens. *Sukiyaki* is a popular cooking tradition that was originally derived from cooking food on a spade held over a fire. From China we could investigate the *yen pi* (fine pastry) of Fukien, stuffed with meat; and the noodle dish of Shantung, *ch'a chiang ming,* like spaghetti with sauce. Shark's fin soup has been highly regarded in China for thousands of years, and it is now believed that shark cartilage is a potent source of antibacterial compounds. *Gingko biloba* is an ancient tree known only from fossils until it was discovered by Western botanists growing in the grounds of ancient Chinese temples. Now, preparations made from gingko are among the most popular herbal supplements sold around the world.

From Vietnam comes *nuoc-mam*, an ancient recipe for fermented fish sauce which can be up to ten years in the making. Cambodia offers *phoat khsat*, royal rice, a dish made from blended chicken, crayfish and pork. Russian *blinis* (buckwheat pancakes) deserve to be more widely known, along with *shashlik*, cubes of marinated lamb seared in the barbecue. From Turkey comes *pastirma*, a form of spiced salt-dried beef, comparable to the pastrami so popular in America.

Indian cuisine has many undiscovered secrets. From the pre-Hindu era comes *n'hari*, a magnificent dish made from curried beef, a reminder of a style of life before the cow became sacred. This rare dish is eaten as a breakfast meal (the name of the dish derives from 'dawn'). Many ancient Indian teachings have been recorded in the Ayur Vedic traditions – perhaps another resource for the kitchen of tomorrow.

Tea tree oil is a new vogue in the West, but has been used by the Aborigines of Australia for its curative properties for thousands of years. Similar products known to the Maoris are being used in New Zealand for the treatment of skin disorders. Taxol, extracted from yew trees in Oregon, is known to act against cancer of the ovary and breast. From the Madagascar rainforest comes the rosy periwinkle *Catharanthus rosea* which has revolutionized the treatment of leukaemia. Many familiar herbs are re-emerging as components of a diet that can heal. *Echinacea*, a popular garden flower, is believed by many enthusiasts to have a beneficial effect on the immune system, and St John's Wort is an increasingly popular herbal product that is claimed to lift the spirits. Care must be taken with all such herbal products. There has been one report of a French chef adding Viagra to the soup but, although the customers liked it, the authorities pounced on him for dispensing a listed drug. This is not the way forward that I envisage, but many conventional spices already in our kitchens are actively psychotropic, and we may well see the development of new food products that cheer the soul in the hard-pressed days that lie ahead.

It is vitally important to standardize these novel dietary supplements. Chinese traditional medicine is much in vogue in the West, as are Western herbal remedies, but herbal products are not always as innocuous as they may seem. There are two major problems with herbal ingredients:

- What herb are we using? Plants known by their popular names are sometimes interpreted differently by different authorities. We need to be sure of the exact identity of the plant, so the Latin name is a prerequisite, and may need to be backed up by genetic sequencing to make doubly sure.
- How much of the vital component does the herb contain? Some plants produce much higher levels of active constituents than other specimens of the same species.

Companies selling pills made from St John's Wort have standardized the strength of the product at 0.3% hypericin. That seems very workable, until you realize that some research concludes that the active ingredient isn't hypericin at all, but hyperforin. Should this become the new standard? Nobody yet knows.

Meanwhile, there remain so many other forms of readily grown protein. In the future we may turn away from beef and poultry, and look to other forms of food, some of which seem strange to the Western world. Rats, snakes, dogs and cats are already consumed as delicacies in some cultures, and grubs, maggots and other fat insect larvae may offer much for everyone the future. We need not imagine that these are only consumed in far-off lands, for in Paris in 1910 the *Société Zoologique d'Acclimatation* served a celebration dinner of Indian python casserole, Algerian tortoise stew and ostrich-egg omelette. Their earlier menus had already featured rat salami and *brochettes* of dog-liver. The *Société* had been established in 1854 to introduce exotic animal novelties into France.

With such a vast range of herbs, plants, fruits and vegetables awaiting discovery by nations who do not know them, we can expect to see a cavalcade of unimaginable new foods for the future. The whole world offers new tastes for tomorrow, from which international cuisine stands to benefit so much. How ironic it seems that many of these countries are currently trying to cultivate a taste for burgers from McDonald's.

Tomorrow's Kitchen

In future, the domestic kitchen will cease to exist in its present form. Once people realize how much of their food is infected with bacteria, kitchen designers will be obliged to incorporate an extra basin for washing the hands in a weak solution of bleach.

With the development of microelectronics, kitchens will be automated, with sensors and regulators able to open and close curtains and control heating dependent upon preset values on a home computer. Bar-coded packs of food will be read by sensors on a refrigerator, and I would like to see food values, calorie counts, and other data made automatically available for the consumer. Remote access can offer far greater opportunities for the busy home manager. It will be possible to call home on a mobile link and set a food-heating program, run the bath, or start the bread-maker.

The labels on food, and the presentation of containers, will be revolutionized. Labelling must become more honest, so that you will immediately know exactly what you buying and consuming. There are products, for example, called 'cranberry drink' and even 'cranberry classic', but they contain only small amounts of the actual juice. One that caught my eye said, in large letters, 'Real Fruit Juice, Cranberry'; but when I looked closely I could read the small print as well. In full, the label said: 'From the Real Fruit Juice Company, Cranberry Flavoured Drink'.

Once the public show an interest in the tricks that copywriters can play, companies will no longer be able to prey on general public ignorance. I would also like to see foods sold, not by weight, but by energy value. Currently, many people (mostly women, but in future the trend will spread to an equal proportion of men) attempt to calculate the energy equivalent of foods, and usually make a bad job of it. In future I'd like you to be able to purchase 100 Calories, or whatever, rather than so many grams or ounces; the recording of daily totals would be much simplified as a result.

The sheer enjoyment of cooking a meal together out of ingredients that are fresh and healthy is a wonderful family occupation. Most children like cookery, but the modern world is becoming increasingly stressful, and it is rare for the family to prepare meals together. The preparation of home-cooked food is already dying out in the West, and this trend will accelerate in the years ahead.

This why I believe that kitchens will disappear from many homes altogether in the future. There will be no wish to cook in the house. Many modern homes already have nowhere to eat, for residents usually snack in front of the television. Eating a communal meal at home will be an unusual event. Food will be heated in a microwave. If there is a cooker, it will be on the dining table, rather than in the kitchen. Communal cookery of food items on skewers, as with the fondue, will replace preparation in the kitchen.

In the West, it is increasingly normal for people to go out for meals, and restaurants are no longer only for special occasions. Some are deliberately aimed at youngsters, with chicken nuggets and hamburgers, but children also like well-prepared adult food, and the fatty and salt-laden foods of today will seem anachronistic in the future.

We may see radical departures from today's conventions as new eating experiences are developed. There will be themed establishments, and meals with entertainment, like a movie or a musical. Mealbars will provide a chance for people to meet and relax.

Finally, along with the scientifically regulated diet of the future, there will be an enthusiasm for the revival of real food – for traditional cookery, freshly grown vegetables and cookery with an emphasis on health. New foods from cross-cultural influences will make eating into an exciting experience.

Even the Arab camelherd Idris, who told me proudly that he had existed all his adult life on camel's milk, and had never eaten food in his life, would revel in the food of tomorrow.

SELECT BIBLIOGRAPHY AND SOURCES

Bittman, Mark, illustrated by Alan Witschonke, *How to Cook Everything : Simple Recipes for Great Food*, Foster City, Calif., 1998

Ford, Brian J., *Microbiology and Food*, London, 1970

—, *The Food Book*, London, 1986

Jaffrey, Madhur, *World Vegetarian*, New York, 1999

McGee, Harold, *On Food and Cooking : The Science and Lore of the Kitchen*, Collier Books, 1997

Macrae, R., R.K. Robinson and M.J. Sadler, *Encyclopaedia of Food Science, Food Technology and Nutrition*, 8 vols, London, San Diego, Sydney and Tokyo, 1993

Mariani, John F., *The Encyclopedia of American Food & Drink*, New York, 1999

Papanek, Victor, *The Green Imperative* , London and New York, 1995

Porritt, Jonathan, *Playing Safe: Science and the Environment*, London and New York, 2000

Rombauer, Irma S., *The New Joy of Cooking*, New York, 1997

Smith, Delia, *How to Cook*, London, 1998

—, *Four Seasons Cookery Book*, London, 1996

Stevens, James Talmage, *Making the Best of Basics: Family Preparedness Handbook*, Salt Lake City, 1997

WEB SITES

http://www.sciences.demon.co.uk
  Author's web site
http://www.nalusda.gov/fnic/ USA
  Agriculture Network Information Center
http://www.foodfuture.org.uk/
  commercial site on biotechnology from the
  UK food industry
http://www.fao.org/
  Food and Agriculture Organisation of the
  United Nations
http://www.thefoodalliance.org/
  alternative critical view on GMOs, etc.
http://207.167.67.197/
  and
http://www.securefuture.com/
  dehydrated food, etc., for emergencies
http://www.rytwayfoods.com/home.cfm
  pre-prepared futuristic food
http://www.west.it/en/1996p02.htm
  and
http://www.vc.attjens.co.jp/VC97_report/
  WEB/VC_01/CFUTUREF.HTM
  students simulate future menus
http://www.lintech.co.uk/
  laboratory developing new flavours, etc.
http://www.britannia.com/science/
  biofood.html
  Britannia Internet magazine on GMOs
http://www.iftf.org/
  Institute for the Future, forecasting
  organization
http://www.exploratorium.edu/genepool/
  scenario_2.html
  public attitude questionnaire
http://www.overpopulation.com/
  future_food.html
  comments on likely global population
http://www.billennium.com/e/eba.htm
  new millennium ideas on food, etc

INDEX